华章IT | HZBOOKS | Information Technology

RocketMQ 技术内幕

RocketMQ架构设计与实现原理

丁威 周继锋◎著

RocketMQ Internals: Architecture, Design and Principle

机械工业出版社
China Machine Press

图书在版编目（CIP）数据

RocketMQ 技术内幕：RocketMQ 架构设计与实现原理 / 丁威，周继锋著. —北京：机械工业出版社，2019.1

ISBN 978-7-111-61421-0

I. R… II. ①丁… ②周… III. 计算机网络 - 软件工具 IV. TP393.07

中国版本图书馆 CIP 数据核字（2018）第 263145 号

RocketMQ 技术内幕

RocketMQ 架构设计与实现原理

出版发行：机械工业出版社（北京市西城区百万庄大街 22 号 邮政编码：100037）	
责任编辑：张锡鹏	责任校对：殷 虹
印　　刷：北京市兆成印刷有限责任公司	版　　次：2019 年 1 月第 1 版第 1 次印刷
开　　本：186mm×240mm 1/16	印　　张：18.75
书　　号：ISBN 978-7-111-61421-0	定　　价：69.00 元

凡购本书，如有缺页、倒页、脱页，由本社发行部调换
客服热线：（010）88379426　88361066　　　投稿热线：（010）88379604
购书热线：（010）68326294　88379649　68995259　读者信箱：hzit@hzbook.com

版权所有 · 侵权必究
封底无防伪标均为盗版
本书法律顾问：北京大成律师事务所　韩光 / 邹晓东

Foreword 推 荐 序

当前，全球商业正面临着产业、组织和价值观三大重构。首先，数字技术重新定义了商业模式，颠覆了原有的产业组织，许多行业中的领头羊，不是被行业内部的追赶者所取代，就是因技术进步而受到挑战，技术变革影响行业竞争格局的速度和深度都发生了数量级上的变化；其次，企业的组织也需要重构，过去的中心、多中心式的组织结构已经不能适应数字时代的需要，新时代的组织要全面向分布式升级。再次，价值观也在重构，数字经济体奉行的价值观是开放、分享、透明和责任，这和传统企业的价值观大相径庭。在这一全新价值观驱动下，企业之间为了更好地推进自身组织结构升级，提高产能，已经将目光逐渐转向开放技术，尤其是云计算领域。而开放、无技术绑定、技术标准恰恰是云计算三个最为典型的特征。在这一领域，RedHat 无疑是冲在了最前头，其最新推出的《开源故事》，由一系列旨在弘扬开源价值（如英才培养、社区构建、透明文化）的精彩影片构成，详细阐述了这一价值正在对我们的工作与生活所产生的影响。从教育领域到制造社区，再到慈善组织和环保事业，无不充分体现了开源价值理念，以及拥抱这一理念的个人如何推动开放文化成为 21 世纪创新的新模式。

人的本性中有一种崇尚自由、希望共享的心理。毕竟，隐私只是我们个人生活的一小部分而已，而我们的社会团体生活中，开放却是促进整个集体进步的重要因素。当我们分清楚什么不能公开和什么该公开的时候，以及当我们很好地平衡这两者而非抑制其中一方的时候，我们的生活就会更加美好。开放它们不但没有使公开者蒙受损失，反而引起了人们更多的兴趣和热情，使得相关的技术发展进入一个良性循环而稳步前进，这就产生了一种良好的社会风气。

RocketMQ 的开源正是源于对这种开源文化的认同，开放是为了更好的协同创新，并将这一技术推向新的高度。在经历了阿里巴巴集团内部多年"双11"交易核心链路工业级场景的验证，2016年11月，团队将RocketMQ捐献给全球享有盛誉的Apache软件基金会，正式成为孵化项目。至此，RocketMQ开启了迈向全球顶级开源软件的新征程。通过社区与团队半年多的开放式创新。2017年9月25日，Apache软件基金会官方宣布，阿里巴巴捐赠给Apache社区的开源分布式消息引擎RocketMQ从社区正式毕业，成为与Apache Spark"同款"的顶级项目（TLP）。最近，我们也看到越来越多的中国本土开源软件进入Apache，Apache优秀的社区理念驱动着更多立志打造世界级品牌的团队不断发展，促进其生态更加健康和活跃。

在过去几年里，RocketMQ在第10届、第16届中北亚开源高峰论坛，以及2017年工信部开源峰会的开源软件评比中，都代表中国更高的开源技术，获得了傲人的开源大奖。在这里，请允许我代表RocketMQ团队，感谢大家这些年对RocketMQ的厚爱与支持。丁威同学是RocketMQ社区里比较早期的的"布道师"，也非常感谢他能将自己这些年的积累整理成书，帮助大家更好地理解RocketMQ，参与到RocketMQ社区建设中来。目前，RocketMQ团队与社区正在构建下一代RocketMQ以及领域标准OpenMessaging，我们希望它是Cloud Native编程范式下首选的金融级高可靠、高性能数据消息计算平台，也非常欢迎大家能够加入到社区建设中来。

人类的生活正在因为开源软件而变得更加美好，让我们一起来构筑美好未来！

冯嘉

Alibaba Messaging 开源技术负责人

Apache RocketMQ 创始人

Linux OpenMessaging 创始人、主席

Preface 前言

为什么要写这本书

随着互联网技术蓬勃发展，微服务架构思想的兴起，系统架构开始追求小型化、轻量化，原有的大型集中式的 IT 系统通常需要进行垂直拆分，孵化出颗粒度更小的众多小型系统，因此对系统间松耦合的要求越来越高，目前 RPC、服务治理、消息中间件几乎成为互联网架构的标配。

引入消息中间件，服务之间可以通过可靠的异步调用，降低系统之间的耦合度，提高系统的可用性。消息中间件的另一个重要应用场景是解决系统之间数据的一致性（最终一致性）。

RocketMQ 作为阿里开源的一款高性能、高吞吐量的消息中间件，承载了阿里"双 11"大部分业务，可以说是一名久经战场的"精英"、值得信任的"伙伴"。同时它的开发语言为 Java，自然而然地得到了广大互联网架构师们的青睐，成为互联网行业首选的消息中间件。

初次接触 RocketMQ 是在听到阿里巴巴正式将 RocketMQ 捐献给 Apache 基金会，成为 Apache 的顶级开源项目时，这意味着承载阿里"双 11"巨大流量的消息中间件完全走向开源，对广大 Java 开发者来说无疑是一个巨大的利好，让我们有机会一睹高性能消息中间件 RocketMQ 的"真容"。作为一名阿里技术崇拜者，我内心异常激动，于是不假思索地在 CSDN 上开通了专栏"源码研究 RocketMQ"，受到了广大技术朋友的支持。

RocketMQ 作为一款高性能消息中间件，其核心优势是可靠的消息存储、消息发送的高性能与低延迟、强大的消息堆积能力与消息处理能力、严格的顺序消息模式等。RocketMQ 的另一个核心思想是懂得取舍。软件设计不可能做到面面俱到，消息中间件的理想状态是一条消息能且只能被消费一次，但要做到这一点，必然需要牺牲性能。RocketMQ 的设计者解决这一难题的办法是不去解决，即保证消息至少被消费一次，但不承诺消息不会被消费者多次消费，其消费的幂等由消费者实现，从而极大地简化了其实现内核，提高了 RocketMQ 的整体性能。

自从 RocketMQ 被捐献给 Apache 基金会后便在快速发展，RocketMQ 的设计者们正在制定消息中间件的新规范，其模块为 openmessaging。本书主要是抛砖引玉，与各位读者朋友们探讨 RocketMQ 的实现原理，使读者能更好地在实际项目中应用 RocketMQ。

读者对象

- RocketMQ 用户和爱好者
- RocketMQ 代码开发志愿者
- Java 中高级开发工程师
- Java 架构师
- 有志于从事 Java 开源的相关技术从业者

本书特色

本书从源码的角度对 RocketMQ 的实现原理进行详细剖析，从中阐述了作者学习阅读源码的方法。本书作为一本源码阅读类书籍，其讲解切入点并不是以组成 RocketMQ 的一个个源码包进行展开，而是基于功能模块（如 Topic 路由中心、消息发送、消息存储、消息消费、事务消息）来展开，更加贴近实战需求。

如何阅读本书

本书分为三大部分。

第一部分为准备篇（第 1 章），简单介绍了 RocketMQ 的设计理念与目标，并介绍了在开发工具中如何对 RocketMQ 进行代码调试。

第二部分为实现篇（第 2~8 章），重点讲解了 RocketMQ 各个功能模块的实现原理，包括 NameServer、消息发送、消息存储、消息消费、消息过滤、顺序消息、事务消息等。

第三部分为实例篇（第 9 章），通过示例展示 RocketMQ 的使用技巧，着重讲解了 RocketMQ 的监控命令与监控管理界面。

本书在最后的附录中给出了 RocketMQ 的主要参数列表及含义，供读者参考。

本书的行文思路主要是根据消息发送的全流程进行展开，从路由管理到消息发送、消息存储、消息消费，再到顺序消息、事务消息，从而实现消息链路的闭环。建议读者按照该思路，带着问题来阅读本书，或许会事半功倍。

勘误和支持

除封面署名外，参加本书编写工作的还有陈鹏飞。由于水平有限，编写时间仓促，书中难免会出现一些错误或者不准确的地方，恳请读者批评指正。为此，大家可以通过CSDN博客专栏（https://blog.csdn.net/column/details/20603.html）留言反馈。书中的全部源文件可以从github rocketmq 官方仓库中下载，我也会将相应的功能及时更新。如果你有更多的宝贵意见，也欢迎发送邮件至 dw19871218pmz@126.com，期待能够得到你的真挚反馈。

致谢

首先要感谢MyCAT开源社区负责人周继锋对我的提携与指导，为我的职业发展指明前进的方向。

感谢RocketMQ联盟中每一位充满创意和活力的朋友——奔腾、zenk、共产国际史派克、水动力皮划艇、张登、张凤凰、曾文、季永超，以及名单之外的很多朋友，感谢你们对我的支持与帮助。感谢杨福川老师的引荐，是你的努力才促成了本书的成功出版。

感谢机械工业出版社华章公司的编辑张锡鹏，在这一年多的时间中始终支持我的写作，你的鼓励和帮助引导我能顺利完成全部书稿。

最后感谢我的爸爸、妈妈、爷爷、奶奶，感谢你们将我培养成人，并时时刻刻为我灌输着信心和力量！感谢我的老婆、女儿，你们是我持续努力的最大动力。

谨以此书献给我最亲爱的家人，以及众多热爱RocketMQ的朋友们！

丁威

感谢RocketMQ团队，是你们的付出才有这么好的产品，同时感谢杨福川编辑对本书出版工作的支持。

谨以此书献给我最亲爱的家人和同事，以及帮助过、关注过我的人，以及使用、学习过RocketMQ的朋友们！

周继锋

目 录 Contents

推荐序
前言

第1章 阅读源代码前的准备 ············ 1

1.1 获取和调试 RocketMQ 的源代码 ··· 1
    1.1.1 Eclipse 获取 RocketMQ 源码 ····· 2
    1.1.2 Eclipse 调试 RocketMQ 源码 ····· 9
    1.1.3 IntelliJ IDEA 获取 RocketMQ
          源码 ·· 15
    1.1.4 IntelliJ IDEA 调试 RocketMQ
          源码 ·· 20
1.2 RocketMQ 源代码的目录结构 ······ 27
1.3 RocketMQ 的设计理念和目标 ······ 28
    1.3.1 设计理念 ······························ 28
    1.3.2 设计目标 ······························ 28

第2章 RocketMQ路由中心 NameServer ························· 31

2.1 NameServer 架构设计 ················ 31
2.2 NameServer 启动流程 ················ 32
2.3 NameServer 路由注册、故障
    剔除 ·· 36
    2.3.1 路由元信息 ·························· 36
    2.3.2 路由注册 ······························ 38
    2.3.3 路由删除 ······························ 43
    2.3.4 路由发现 ······························ 46
2.4 本章小结 ···································· 47

第3章 RocketMQ消息发送 ············ 49

3.1 漫谈 RocketMQ 消息发送 ·········· 49
3.2 认识 RocketMQ 消息 ·················· 50
3.3 生产者启动流程 ························· 51
    3.3.1 初识 DefaultMQProducer
          消息发送者 ··························· 51
    3.3.2 消息生产者启动流程 ··········· 54
3.4 消息发送基本流程 ····················· 56
    3.4.1 消息长度验证 ······················ 56
    3.4.2 查找主题路由信息 ··············· 56
    3.4.3 选择消息队列 ······················ 60
    3.4.4 消息发送 ······························ 65
3.5 批量消息发送 ···························· 71
3.6 本章小结 ···································· 74

第4章 RocketMQ消息存储 ············ 75

4.1 存储概要设计 ···························· 75

4.2 初识消息存储 ······ 76
4.3 消息发送存储流程 ······ 78
4.4 存储文件组织与内存映射 ······ 83
    4.4.1 MappedFileQueue 映射文件队列 ······ 84
    4.4.2 MappedFile 内存映射文件 ······ 87
    4.4.3 TransientStorePool ······ 93
4.5 RocketMQ 存储文件 ······ 94
    4.5.1 Commitlog 文件 ······ 95
    4.5.2 ConsumeQueue 文件 ······ 97
    4.5.3 Index 索引文件 ······ 100
    4.5.4 checkpoint 文件 ······ 104
4.6 实时更新消息消费队列与索引文件 ······ 105
    4.6.1 根据消息更新 ConumeQueue ······ 107
    4.6.2 根据消息更新 Index 索引文件 ······ 108
4.7 消息队列与索引文件恢复 ······ 109
    4.7.1 Broker 正常停止文件恢复 ······ 112
    4.7.2 Broker 异常停止文件恢复 ······ 114
4.8 文件刷盘机制 ······ 115
    4.8.1 Broker 同步刷盘 ······ 116
    4.8.2 Broker 异步刷盘 ······ 119
4.9 过期文件删除机制 ······ 122
4.10 本章小结 ······ 126

第5章 RocketMQ消息消费 ······ 127

5.1 RocketMQ 消息消费概述 ······ 127
5.2 消息消费者初探 ······ 128
5.3 消费者启动流程 ······ 130

5.4 消息拉取 ······ 133
    5.4.1 PullMessageService 实现机制 ······ 133
    5.4.2 ProcessQueue 实现机制 ······ 136
    5.4.3 消息拉取基本流程 ······ 138
5.5 消息队列负载与重新分布机制 ······ 154
5.6 消息消费过程 ······ 162
    5.6.1 消息消费 ······ 163
    5.6.2 消息确认 (ACK) ······ 167
    5.6.3 消费进度管理 ······ 171
5.7 定时消息机制 ······ 176
    5.7.1 load 方法 ······ 177
    5.7.2 start 方法 ······ 178
    5.7.3 定时调度逻辑 ······ 179
5.8 消息过滤机制 ······ 181
5.9 顺序消息 ······ 186
    5.9.1 消息队列负载 ······ 187
    5.9.2 消息拉取 ······ 187
    5.9.3 消息消费 ······ 188
    5.9.4 消息队列锁实现 ······ 195
5.10 本章小结 ······ 196

第6章 消息过滤FilterServer ······ 198

6.1 ClassFilter 运行机制 ······ 198
6.2 FilterServer 注册剖析 ······ 199
6.3 类过滤模式订阅机制 ······ 202
6.4 消息拉取 ······ 205
6.5 本章小结 ······ 206

第7章 RocketMQ主从同步(HA)机制 ······ 207

- 7.1 RocketMQ 主从复制原理 ········ 207
  - 7.1.1 HAService 整体工作机制 ····· 208
  - 7.1.2 AcceptSocketService 实现原理 ······ 208
  - 7.1.3 GroupTransferService 实现原理 ······ 210
  - 7.1.4 HAClient 实现原理 ············ 211
  - 7.1.5 HAConnection 实现原理 ····· 214
- 7.2 RocketMQ 读写分离机制 ········ 220
- 7.3 本章小结 ······ 223

第8章 RocketMQ事务消息 ······ 225

- 8.1 事务消息实现思想 ······ 225
- 8.2 事务消息发送流程 ······ 226
- 8.3 提交或回滚事务 ······ 232
- 8.4 事务消息回查事务状态 ······ 233
- 8.5 本章小结 ······ 240

第9章 RocketMQ实战 ······ 242

- 9.1 消息批量发送 ······ 242
- 9.2 消息发送队列自选择 ······ 243
- 9.3 消息过滤 ······ 243
  - 9.3.1 TAG 模式过滤 ······ 244
  - 9.3.2 SQL 表达模式过滤 ······ 244
  - 9.3.3 类过滤模式 ······ 245
- 9.4 事务消息 ······ 247
- 9.5 Spring 整合 RocketMQ ······ 250
- 9.6 Spring Cloud 整合 RocketMQ ···· 251
- 9.7 RocketMQ 监控与运维命令 ······ 258
  - 9.7.1 RocktetMQ 监控平台搭建 ···· 258
  - 9.7.2 RocketMQ 管理命令 ······ 261
- 9.8 应用场景分析 ······ 280
- 9.9 本章小结 ······ 281

附录A 参数说明 ······ 282

第 1 章

阅读源代码前的准备

研究一款开源中间件，首先我们需要了解它的整体架构以及如何在开发环境调试源码，从代码入手才能快速熟悉一个开源项目，只有这样才能抽丝剥茧地理解透彻，了解作者的设计思想和实现原理。本章将重点介绍 RocketMQ 的整体设计理念以及如何调试 RocketMQ，为后续源码阅读打下扎实的基础。

本章重点内容如下。
- 获取和调试 RocketMQ 源代码
- RocketMQ 源代码的目录结构
- RocketMQ 的设计理念和设计目标

1.1　获取和调试 RocketMQ 的源代码

RocketMQ 原先是阿里巴巴内部使用的消息中间件，于 2017 年提交到 Apache 基金会成为 Apache 基金会的顶级开源项目，GitHub 代码库链接：https://github.com/apache/rocketmq.git。在 Github 网站上搜索 RocketMQ，如图 1-1 所示。

2 ❖ RocketMQ 技术内幕：RocketMQ 架构设计与实现原理

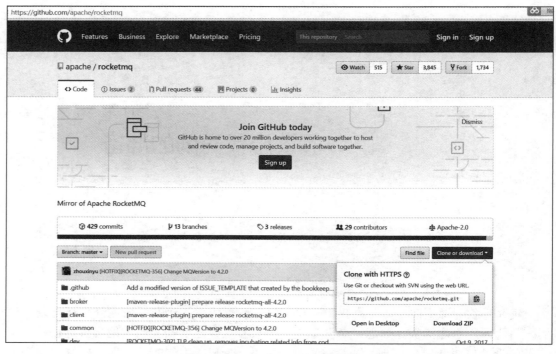

图 1-1　GitHub RocketMQ 搜索界面

1.1.1　Eclipse 获取 RocketMQ 源码

Step1：单击右键从菜单中选择 import git，弹出如图 1-2 所示的对话框。

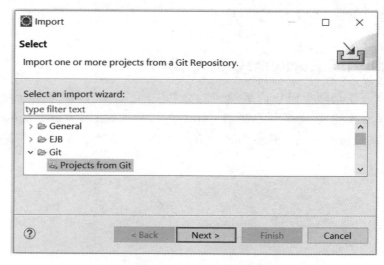

图 1-2　Import 对话框

第 1 章　阅读源代码前的准备　◆◆　3

Step2：点击 Next 按钮，弹出 Projects from Git 对话框，如图 1-3 所示。

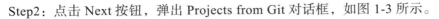

图 1-3　Import Projects from Git 对话框

Step3：点击 Next 按钮，弹出 Clone URI 对话框，如图 1-4 所示。

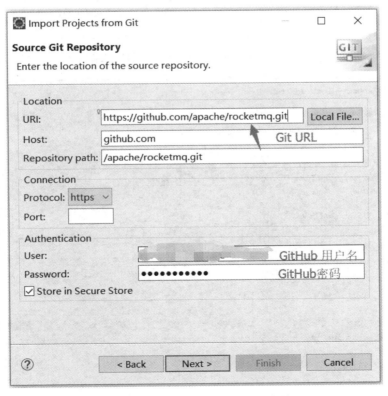

图 1-4　Import Projects from Git 对话框

Step4：继续点击 Next 进入下一步，选择代码分支，如图 1-5 所示。

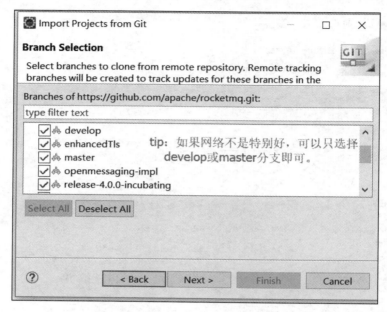

图 1-5　Import Projects from Git 对话框

Step5：选择所需要的分支后点击 Next，进入代码存放目录选择，如图 1-6 所示。

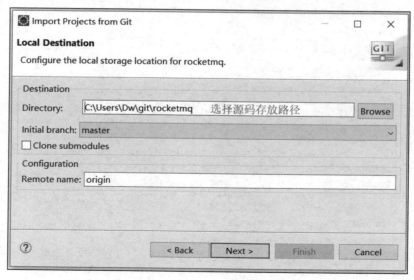

图 1-6　Import Projects from Git 对话框

Step6：点击 Next，Eclipse 将从远程仓库下载代码，如图 1-7 所示。

第 1 章 阅读源代码前的准备 5

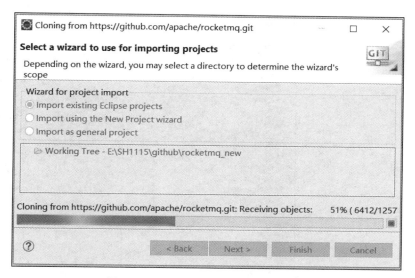

图 1-7 Import Projects from Git 对话框

Step7：代码下载到指定目录后，默认选择 Import existing projects（单分支），这里手动选择 Import as general projects（多分支），点击 Finish，成功导入，如图 1-8 所示。

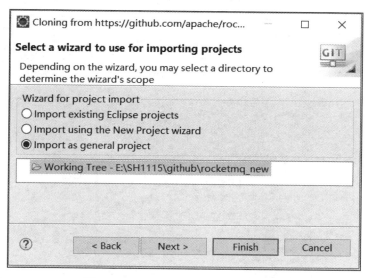

图 1-8 Import Projects from Git 对话框

Step8：代码导入成功后，需要将项目转换成 Maven 项目，导入成功后的效果图，如图 1-9 所示。

Step9：单击右键从上下文菜单中选择 rocketmq_new（文件下载目录名）→ Configure → Convert and Detect Nested Projects 转换成 Maven 项目，如图 1-10 所示。

6 ❖ RocketMQ 技术内幕：RocketMQ 架构设计与实现原理

图 1-9　导入项目初始状态

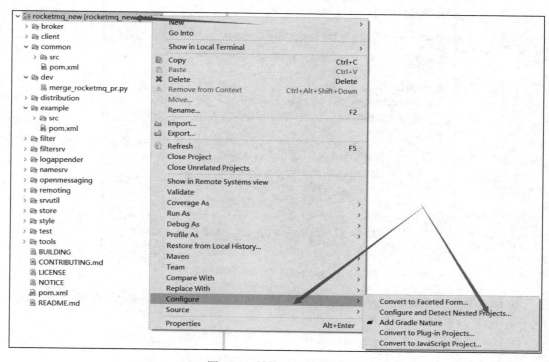

图 1-10　转换 Maven 项目

Step10：点击 Finish 执行 Maven 项目转换，完成 RocketMQ 的导入，如图 1-11 所示。

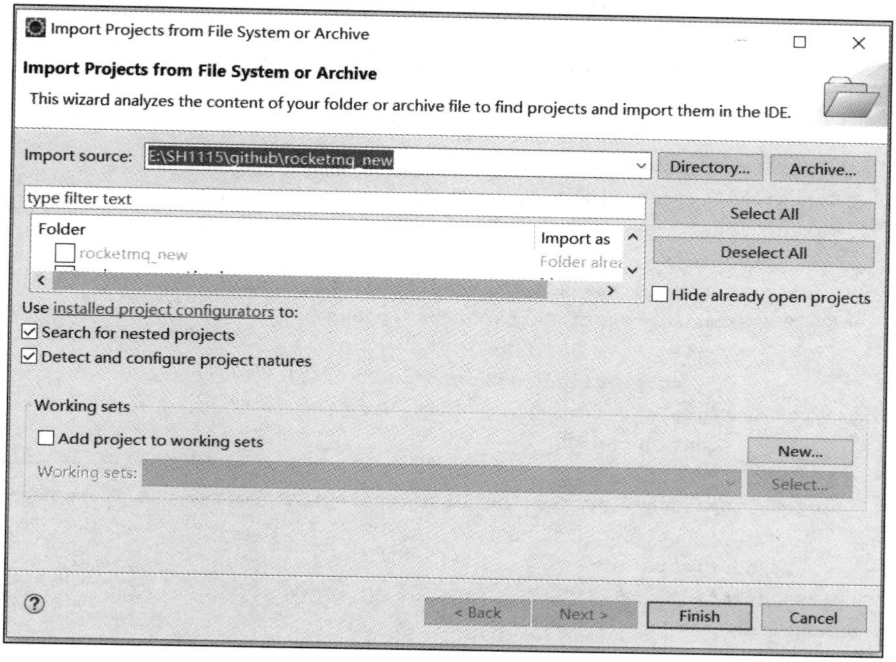

图 1-11　转换 Maven 项目

转换过程中可能会弹出如图 1-12 所示提示框。

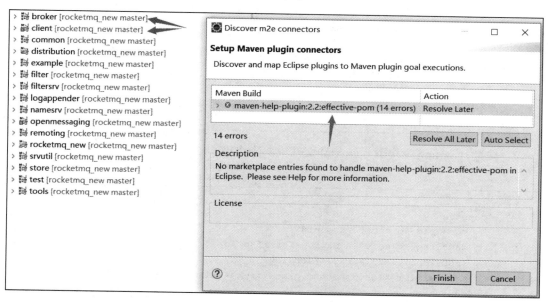

图 1-12　转换 Maven 项目

解决办法有三种。

1）修改根 pom.xml 文件，找到如下条目，加上注释。

代码清单 1-1　rocketmq 根 pom.xml 文件

```xml
<!--
    <plugin>
        <artifactId>maven-help-plugin</artifactId>
        <version>2.2</version>
        <executions>
            <execution>
                <id>generate-effective-dependencies-pom</id>
                <phase>generate-resources</phase>
                <goals>
                    <goal>effective-pom</goal>
                </goals>
                <configuration>

                    <output>${project.build.directory}/effective-pom/effective-depende
                        ncies.xml</output>
                </configuration>
            </execution>
        </executions>
    </plugin>
-->
<!--
    <plugin>
        <artifactId>maven-surefire-plugin</artifactId>
        <version>2.19.1</version>
        <configuration>
            <forkCount>1</forkCount>
            <reuseForks>true</reuseForks>
        </configuration>
    </plugin>
-->
```

2）注释 remoting 模块下 pom.xml 文件中部分代码。

代码清单 1-2　rocketmq 根 pom.xml 文件

```xml
<!--
    <dependency>
        <groupId>io.netty</groupId>
        <artifactId>netty-tcnative</artifactId>
        <version>1.1.33.Fork22</version>
        <classifier>${os.detected.classifier}</classifier>
    </dependency>
-->
```

3）右键一个项目，选择 Maven → Update Project，如图 1-13 所示。

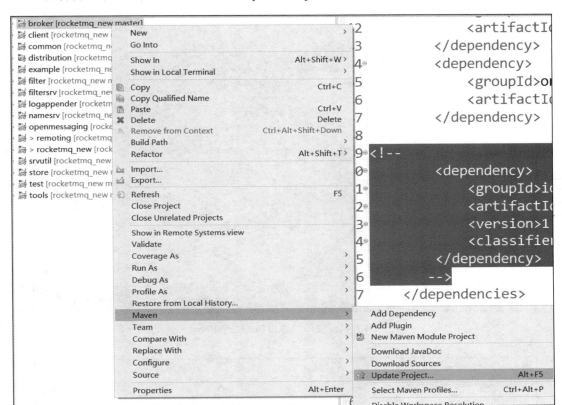

图 1-13　更新 Maven 项目

1.1.2　Eclipse 调试 RocketMQ 源码

本节将展示在 Eclipse 中启动 NameServer、Broker，并运行消息发送与消息消费示例程序。

1. 启动 NameServer

Step1：展开 namesrv 模块，右键 NamesrvStartup.java，移动到 Debug As，选中 Debug Configurations，弹出 Debug Configurations 对话框，如图 1-14 所示。

Step2：选中 Java Application 条目并单击右键，选择 New 弹出 Debug Configurations 对话框，如图 1-15 所示。

Step3：设置 RocketMQ 运行主目录。选择 Environment 选项卡，添加环境变量 ROCKET_HOME。

Step4：在 RocketMQ 运行主目录中创建 conf、logs、store 三个文件夹，如图 1-16 所示。

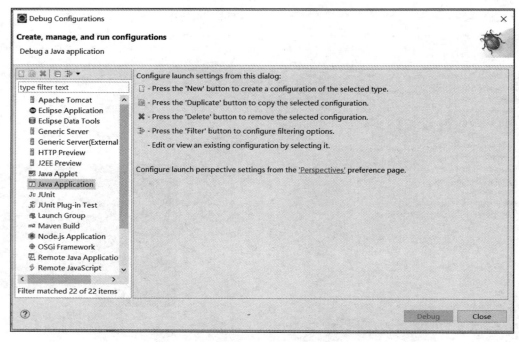

图 1-14　选择 Debug Configurations

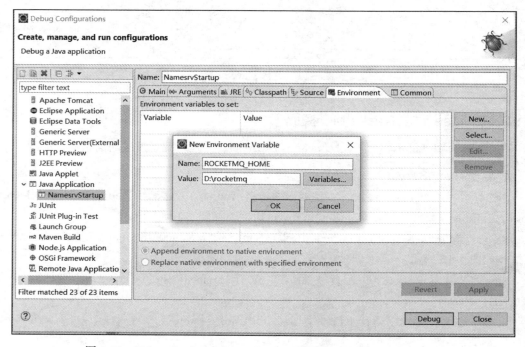

图 1-15　Debug Configurations，Create, manage, and run configurations

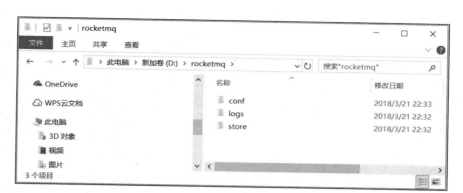

图 1-16 RocketMQ 主目录

Step5：从 RocketMQ distribution 部署目录中将 broker.conf、logback_broker.xml 文件复制到 conf 目录中，logback_namesrv.xml 文件则只需修改日志文件的目录，broker.conf 文件内容如下所示。

代码清单 1-3　broker.conf 文件

```
brokerClusterName=DefaultCluster
brokerName=broker-a
brokerId=0
#nameServer 地址，分号分割
namesrvAddr=127.0.0.1:9876
deleteWhen=04
fileReservedTime=48
brokerRole=ASYNC_MASTER
flushDiskType=ASYNC_FLUSH
# 存储路径
storePathRootDir=D:\\rocketmq\\store
#commitLog 存储路径
storePathCommitLog=D:\\rocketmq\\store\\commitlog
# 消费队列存储路径
storePathConsumeQueue=D:\\rocketmq\\store\\consumequeue
# 消息索引存储路径
storePathIndex=D:\\rocketmq\\store\\index
#checkpoint 文件存储路径
storeCheckpoint=D:\\rocketmq\\store\\checkpoint
#abort 文件存储路径
abortFile=D:\\rocketmq\\store\\abort
```

Step6：在 Eclipse Debug 中运行 NamesrvStartup，并输出 "The Name Server boot success. Serializetype=JSON"。

2. 启动 Broker

Step1：展开 broker 模块，右键 BrokerStartup.java，移动到 Debug As，选中 Debug Configurations，弹出如图 1-17 所示的对话框，选择 arguments 选项卡，配置 -c 属性指定 broker 配置文件路径。

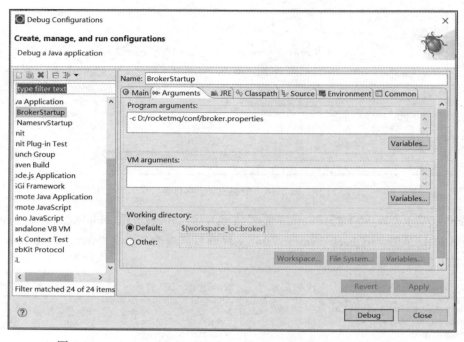

图 1-17　Debug Configurations，Create,manage,and run configurations

Step2：切换选项卡 Environment，配置 RocketMQ 主目录，如图 1-18 所示。

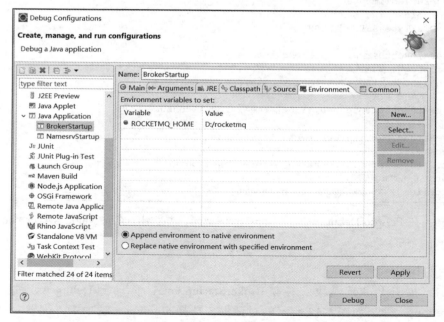

图 1-18　Debug Configurations，Create,manage,and run configurations

Step3：以 Debug 模式运行 BrokerStartup.java，查看 ${ROCKET_HOME}/logs/broker.log 文件，未报错则表示启动成功。

代码清单 1-4　broker 启动日志截图

```
2018-03-22 20:47:29 INFO main - register broker to name server 127.0.0.1:9876 OK
2018-03-22 20:47:29 INFO main - The broker[broker-a, 192.168.1.3:10911] boot success. serializeType=JSON and name server is 127.0.0.1:9876
2018-03-22 20:47:38 INFO BrokerControllerScheduledThread1 - dispatch behind commit log 0 bytes
2018-03-22 20:47:38 INFO BrokerControllerScheduledThread1 - Slave fall behind master: 0 bytes
2018-03-22 20:47:39 INFO BrokerControllerScheduledThread1 - register broker to name server 127.0.0.1:9876 OK
2018-03-22 20:48:09 INFO BrokerControllerScheduledThread1 - register broker to name server 127.0.0.1:9876 OK
2018-03-22 20:48:37 INFO BrokerControllerScheduledThread1 - dispatch behind commit log 0 bytes
2018-03-22 20:48:37 INFO BrokerControllerScheduledThread1 - Slave fall behind master: 0 bytes
2018-03-22 20:48:39 INFO BrokerControllerScheduledThread1 - register broker to name server 127.0.0.1:9876 OK
2018-03-22 20:49:09 INFO BrokerControllerScheduledThread1 - register broker to name server 127.0.0.1:9876 OK
```

3. 使用 RocketMQ 提供的实例验证消息发送与消息消费

Step1：修改 org.apache.rocketmq.example.quickstart.Producer 示例程序，设置消息生产者 NameServer 地址。

代码清单 1-5　消息发送示例程序

```java
public class Producer {
    public static void main(String[] args) throws MQClientException,
                    InterruptedException {
        DefaultMQProducer producer = new
                    DefaultMQProducer("please_rename_unique_group_name");
        producer.setNamesrvAddr("127.0.0.1:9876");
        producer.start();
        for (int i = 0; i < 1; i++) {
            try {
                Message msg = new Message("TopicTest"/* Topic */,"TagA"/* Tag */,
                    ("Hello RocketMQ " + i).getBytes
                        (RemotingHelper.DEFAULT_CHARSET)/* Message body */
                );
                SendResult sendResult = producer.send(msg);
                System.out.printf("%s%n", sendResult);
            } catch (Exception e) {
                e.printStackTrace();
                Thread.sleep(1000);
            }
        }
        producer.shutdown();
```

```
    }
}
```

Step2：运行该示例程序，查看运行结果，如果输出代码清单 1-6 所示结果则表示消息发送成功。

代码清单 1-6　消息发送结果

```
SendResult [sendStatus=SEND_OK, msgId=C0A8010325B46D06D69C70A211400000,
offsetMsgId=C0A8010300002A9F0000000000000000, messageQueue=MessageQueue
[topic=TopicTest, brokerName=broker-a, queueId=0], queueOffset=0]
```

Step3：修改 org.apache.rocketmq.example.quickstart.Consumer 示例程序，设置消息消费者 NameServer 地址。

代码清单 1-7　消息消费示例程序

```
public class Consumer {
    public static void main(String[] args) throws InterruptedException,
        MQClientException {
        DefaultMQPushConsumer consumer = new
            DefaultMQPushConsumer("please_rename_unique_group_name_4");
        consumer.setNamesrvAddr("127.0.0.1:9876");
        consumer.setConsumeFromWhere(ConsumeFromWhere.CONSUME_FROM_FIRST_OFFSET);
        consumer.subscribe("TopicTest", "*");
        consumer.registerMessageListener(new MessageListenerConcurrently() {
            public ConsumeConcurrentlyStatus consumeMessage(List<MessageExt> msgs,
                ConsumeConcurrentlyContext context) {
                System.out.printf("%s Receive New Messages: %s %n",
                    Thread.currentThread().getName(), msgs);
                return ConsumeConcurrentlyStatus.CONSUME_SUCCESS;
            }
        });
        consumer.start();
        System.out.printf("Consumer Started.%n");
    }
}
```

Step4：运行消息消费者程序，如果输出如下所示则表示消息消费成功。

代码清单 1-8　消息消费结果

```
Consumer Started.
ConsumeMessageThread_1 Receive New Messages: [MessageExt [queueId=0,
storeSize=178, queueOffset=0, sysFlag=0, bornTimestamp=1521723269443,
bornHost=/192.168.1.3:57034, storeTimestamp=1521723269510,
storeHost=/192.168.1.3:10911, msgId=C0A8010300002A9F0000000000000000,
commitLogOffset=0, bodyCRC=613185359, reconsumeTimes=0,
preparedTransactionOffset=0, toString()=Message [topic=TopicTest, flag=0,
properties={MIN_OFFSET=0, MAX_OFFSET=1, CONSUME_START_TIME=1521723841419,
UNIQ_KEY=C0A8010325B46D06D69C70A211400000, WAIT=true, TAGS=TagA}, body=16]]]
```

消息发送与消息消费都成功,则说明 RocketMQ 调试环境已经成功搭建了,可以直接 Debug 源码,探知 RocketMQ 的实现奥秘了。

1.1.3 IntelliJ IDEA 获取 RocketMQ 源码

Step1:在 IntelliJ IDEA VCS 菜单中选择 Check from Version Control,再选择 Git,如图 1-19 所示。

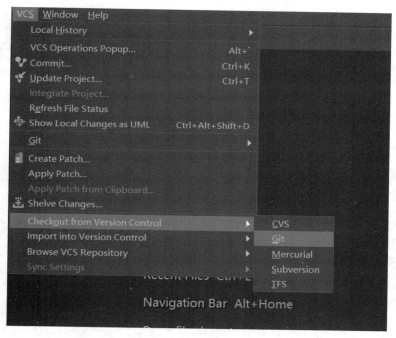

图 1-19 Git clone 对话框

Step2:在弹出的对话框中,URL 输入 RocketMQ 源码地址,选择保存的本地路径,点击 Clone,弹出 Git Repository 对话框,如图 1-20 所示。

图 1-20 Git Repository 对话框

状态栏有代码检出的进度，如图 1-21 所示。

图 1-21　RocketMQ Clon 进度条

Step3：检出完成会弹出提示框，选择 Yes，如图 1-22 所示。

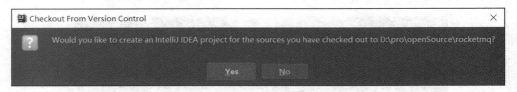

图 1-22　Checkout From Version Control

Step4：在弹出框中选择 Maven，点击 Next，如图 1-23 所示。

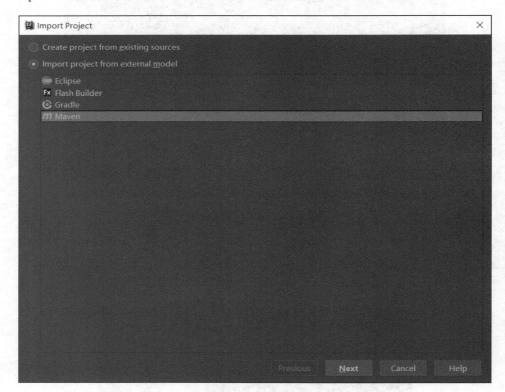

图 1-23　Import Project

Step5：勾选 jdk8，点击 Next，如图 1-24 所示。
Step6：下面 2 步都直接点击 Next，如图 1-25、图 1-26 和图 1-27 所示。

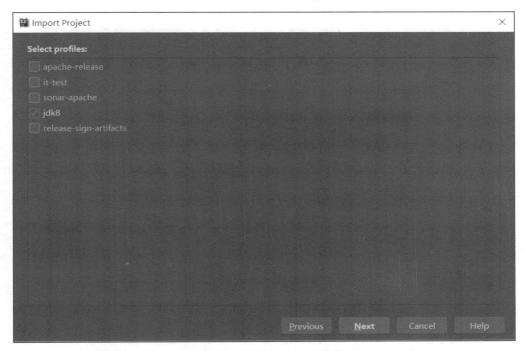

图 1-24　Import Project select profiles

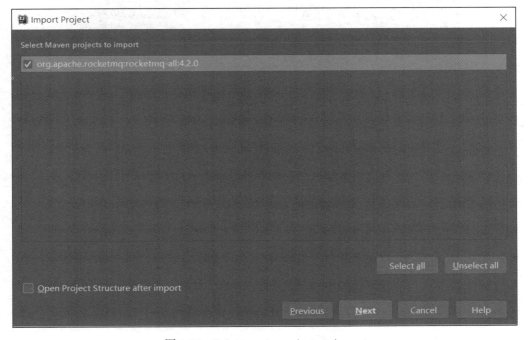

图 1-25　Select maven projects to import

18 ❖ RocketMQ 技术内幕：RocketMQ 架构设计与实现原理

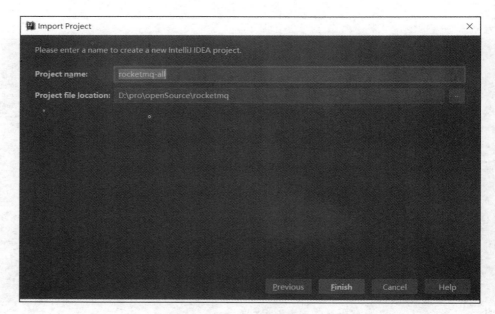

图 1-26 Create a new Project

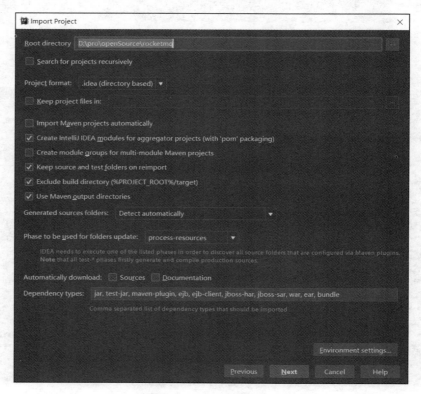

图 1-27 Import RocketMQ

Step7：导入成功后，效果图如图 1-28 所示。

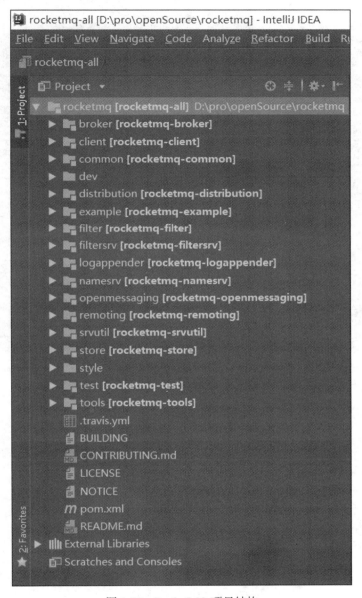

图 1-28　RocketMQ 项目结构

Step8：执行 maven 命令 clean install，进行编译和下载依赖，下载完成后，可以看到控制台 BUILD SUCCESS 的提示信息，如图 1-29 所示。

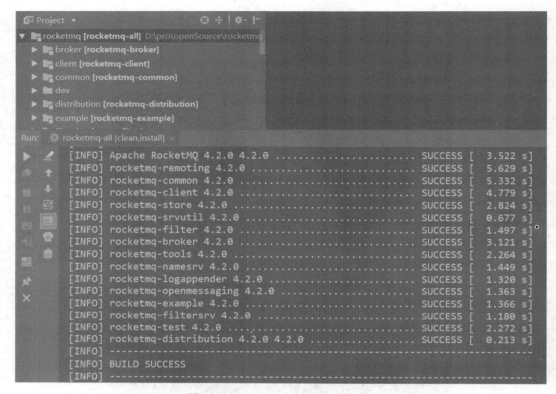

图 1-29　mvn clean install RocketMQ

1.1.4　IntelliJ IDEA 调试 RocketMQ 源码

本节将展示在 IntelliJ IDEA 中启动 NameServer、Broker，并运行消息发送与消息消费示例程序。

1. 启动 NameServer

Step1：展开 namesrv 模块，右键 NamesrvStartup.java，移动到 Debug As，选中 Debug 'NamesrvStartup.java.main()'，弹出如图 1-30、图 1-31 所示的对话框。

Step2：点击 Environment variables 后面的按钮，弹出 Environment variables 对话框，如图 1-32 所示。

Step3：点击"+"号，在 Name 输入框中输入 ROCKETMQ_HOME，Value 输入源码的保存路径。点击 OK，回到 Debug Configurations 界面，再点击 OK，如图 1-33 所示。

Step4：在 RocketMQ 运行主目录中创建 conf、logs、store 三个文件夹。

第 1 章　阅读源代码前的准备　❖　21

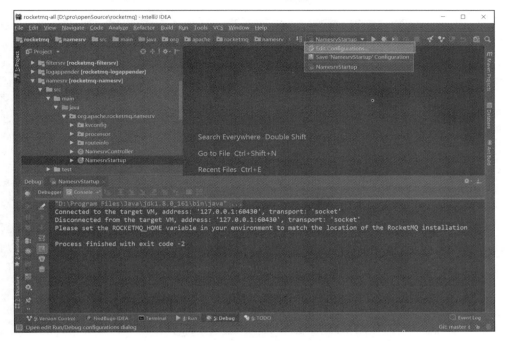

图 1-30　NamesrvStartup Debug

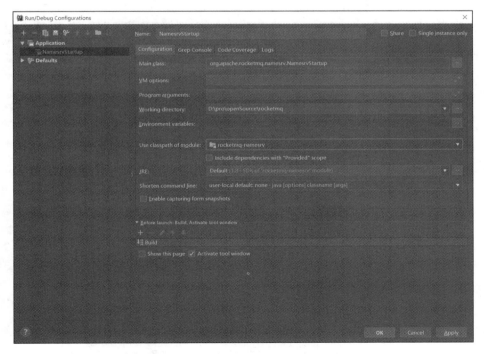

图 1-31　NamesrvStartup Debug Configurations

图 1-32　Environment Variables 列表

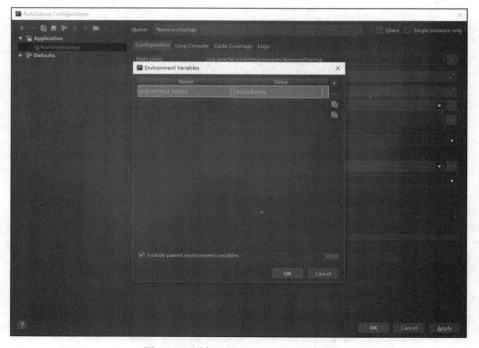

图 1-33　增加 Rocket home 环境变量

Step5：从 RocketMQ distribution 部署目录中将 broker.conf、logback_broker.xml 文件复制到 conf 目录中，logback_namesrv.xml 文件，只需修改日志文件的目录，broker.conf 文件目录内容代码清单 1-9 所示。

代码清单 1-9　broker.conf 文件

```
brokerClusterName=DefaultCluster
brokerName=broker-a
brokerId=0
#nameServer 地址，分号分割
namesrvAddr=127.0.0.1:9876
deleteWhen=04
fileReservedTime=48
brokerRole=ASYNC_MASTER
flushDiskType=ASYNC_FLUSH
# 存储路径
storePathRootDir=D:\\rocketmq\\store
#commitLog 存储路径
storePathCommitLog=D:\\rocketmq\\store\\commitlog
# 消费队列存储路径
storePathConsumeQueue=D:\\rocketmq\\store\\consumequeue
# 消息索引存储路径
storePathIndex=D:\\rocketmq\\store\\index
#checkpoint 文件存储路径
storeCheckpoint=D:\\rocketmq\\store\\checkpoint
#abort 文件存储路径
abortFile=D:\\rocketmq\\store\\abort
```

Step6：在 IntelliJ IDEA Debug 中运行 NamesrvStartup，并输出 "The Name Server boot success. Serializetype=JSON"。

2. 启动 Broker

Step1：展开 broker 模块，右键 BrokerStartup.java 执行，会提示需要配置 ROCKETMQ_HOME。在 idea 右上角选中 Debug Configurations，在弹出的对话框中选择 arguments 选项卡，配置 -c 属性指定 broker 配置文件路径，如图 1-34 所示。

Step2：切换选项卡 Environment，配置 RocketMQ 主目录和 broker 配

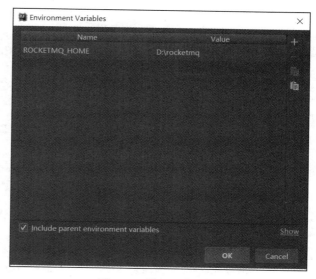

图 1-34　设置环境变量

置文件,如图 1-35 所示。

图 1-35 运行或调试运行时的环境设置

Step3:以 Debug 模式运行 BrokerStartup.java,查看 ${ROCKET_HOME}/logs/broker.log 文件,未报错则表示启动成功,如代码清单 1-10 所示。

代码清单 1-10　broker 启动日志截图

```
    2018-06-15 17:14:27 INFO PullRequestHoldService - PullRequestHoldService
service started
    2018-06-15 17:14:28 INFO main - register broker to name server 127.0.0.1:9876
OK
    2018-06-15 17:14:28 INFO main - The broker[broker-a, 192.168.41.1:10911] boot
success. serializeType=JSON and name server is 127.0.0.1:9876
    2018-06-15 17:14:37 INFO BrokerControllerScheduledThread1 - dispatch behind
commit log 0 bytes
    2018-06-15 17:14:37 INFO BrokerControllerScheduledThread1 - Slave fall behind
master: 534 bytes
    2018-06-15 17:14:38 INFO BrokerControllerScheduledThread1 - register broker to
name server 127.0.0.1:9876 OK
    2018-06-15 17:14:41 INFO ClientManageThread_1 - new consumer connected, group:
please_rename_unique_group_name_4 CONSUME_PASSIVELY CLUSTERING channel:
```

```
ClientChannelInfo [channel=[id: 0x5babb0b1, L:/192.168.41.1:10911 -
R:/192.168.41.1:50635], clientId=192.168.41.1@15140, language=JAVA,
version=253, lastUpdateTimestamp=1529054081078]
    2018-06-15 17:14:41 INFO ClientManageThread_1 - subscription changed, add new
topic, group: please_rename_unique_group_name_4 SubscriptionData
[classFilterMode=false, topic=%RETRY%please_rename_unique_group_name_4,
subString=*, tagsSet=[], codeSet=[], subVersion=1529053720311,
expressionType=null]
    2018-06-15 17:14:41 INFO ClientManageThread_1 - subscription changed, add new
topic, group: please_rename_unique_group_name_4 SubscriptionData
[classFilterMode=false, topic=TopicTest, subString=*, tagsSet=[], codeSet=[],
subVersion=1529053720326, expressionType=null]
    2018-06-15 17:14:41 INFO ClientManageThread_1 - registerConsumer info changed
ConsumerData [groupName=please_rename_unique_group_name_4,
consumeType=CONSUME_PASSIVELY, messageModel=CLUSTERING,
consumeFromWhere=CONSUME_FROM_FIRST_OFFSET, unitMode=false,
subscriptionDataSet=[SubscriptionData [classFilterMode=false,
topic=%RETRY%please_rename_unique_group_name_4, subString=*, tagsSet=[],
codeSet=[], subVersion=1529053720311, expressionType=null], SubscriptionData
[classFilterMode=false, topic=TopicTest, subString=*, tagsSet=[], codeSet=[],
subVersion=1529053720326, expressionType=null]]] 192.168.41.1:50635
    2018-06-15 17:14:41 INFO ClientManageThread_1 - new producer connected, group:
CLIENT_INNER_PRODUCER channel: ClientChannelInfo [channel=[id: 0x5babb0b1,
L:/192.168.41.1:10911 - R:/192.168.41.1:50635], clientId=192.168.41.1@15140,
language=JAVA, version=253, lastUpdateTimestamp=1529054081079]
```

3. 使用 RocketMQ 提供的实例验证消息发送与消息消费

Step1：修改 org.apache.rocketmq.example.quickstart.Producer 示例程序，设置消息生产者 NameServer 地址。

代码清单 1-11　消息发送示例程序

```java
public class Producer {
    public static void main(String[] args) throws MQClientException,
            InterruptedException {
        DefaultMQProducer producer = new
                DefaultMQProducer("please_rename_unique_group_name");
        producer.setNamesrvAddr("127.0.0.1:9876");
        producer.start();
        for (int i = 0; i < 1; i++) {
            try {
                Message msg = new Message("TopicTest" /* Topic */,"TagA" /* Tag */,
                    ("Hello RocketMQ " + i).getBytes
                        (RemotingHelper.DEFAULT_CHARSET)/* Message body */
                );
                SendResult sendResult = producer.send(msg);
                System.out.printf("%s%n", sendResult);
            } catch (Exception e) {
                e.printStackTrace();
                Thread.sleep(1000);
```

```
        }
    }
    producer.shutdown();
}
```

Step2：运行该示例程序，查看运行结果，如果输出如下所示则表示消息发送成功。

代码清单 1-12　消息发送结果

```
SendResult [sendStatus=SEND_OK, msgId=C0A8006606EC18B4AAC24BC584450000,
offsetMsgId=C0A8290100002A9F00000000000000B2, messageQueue=MessageQueue
[topic=TopicTest, brokerName=broker-a, queueId=3], queueOffset=0]
```

Step3：修改 org.apache.rocketmq.example.quickstart.Consumer 示例程序，设置消息消费者 NameServer 地址。

代码清单 1-13　消息消费示例程序

```
public class Consumer {
    public static void main(String[] args) throws InterruptedException,
            MQClientException {
        DefaultMQPushConsumer consumer = new
            DefaultMQPushConsumer("please_rename_unique_group_name_4");
        consumer.setNamesrvAddr("127.0.0.1:9876");
        consumer.setConsumeFromWhere(ConsumeFromWhere.CONSUME_FROM_FIRST_OFFSET);
        consumer.subscribe("TopicTest", "*");
        consumer.registerMessageListener(new MessageListenerConcurrently() {
            public ConsumeConcurrentlyStatus consumeMessage(List<MessageExt> msgs,
                ConsumeConcurrentlyContext context) {
                System.out.printf("%s Receive New Messages: %s %n",
                    Thread.currentThread().getName(), msgs);
                return ConsumeConcurrentlyStatus.CONSUME_SUCCESS;
            }
        });
        consumer.start();
        System.out.printf("Consumer Started.%n");
    }
}
```

Step4：运行消息消费者程序，如果输出如下所示表示消息消费成功。

代码清单 1-14　消息消费结果

```
Consumer Started.
ConsumeMessageThread_1 Receive New Messages: [MessageExt [queueId=0,
storeSize=178, queueOffset=1, sysFlag=0, bornTimestamp=1529053736201,
bornHost=/192.168.41.1:50331, storeTimestamp=1529053736210,
storeHost=/192.168.41.1:10911, msgId=C0A8290100002A9F0000000000000164,
commitLogOffset=356, bodyCRC=613185359, reconsumeTimes=0,
preparedTransactionOffset=0, toString()=Message [topic=TopicTest, flag=0,
```

```
properties={MIN_OFFSET=0, MAX_OFFSET=2, CONSUME_START_TIME=1529053736226,
UNIQ_KEY=C0A800662C8C18B4AAC24BC70D080000, WAIT=true, TAGS=TagA}, body=16]]]
```

消息发送与消息消费都成功,则说明 RocketMQ 调试环境已成功搭建,可以通过 Debug 调试源码,探知 RocketMQ 的实现奥秘了。

1.2 RocketMQ 源代码的目录结构

RocketMQ 源码组织方式基于 Maven 按模块组织,如图 1-36 所示。

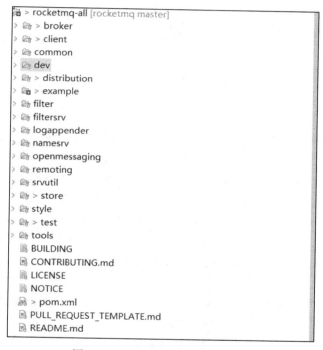

图 1-36　RocketMQ 源码目录结构

RocketMQ 核心目录说明如下。

1) broker:broker 模块 (broker 启动进程)。
2) client:消息客户端,包含消息生产者、消息消费者相关类。
3) common:公共包。
4) dev:开发者信息 (非源代码)。
5) distribution:部署实例文件夹 (非源代码)。
6) example:RocketMQ 示例代码。
7) filter:消息过滤相关基础类。

8）filtersrv：消息过滤服务器实现相关类 (Filter 启动进程)。

9）logappender：日志实现相关类。

10）namesrv：NameServer 实现相关类 (NameServer 启动进程)。

11）openmessaging：消息开放标准，正在制定中。

12）remoting：远程通信模块，基于 Netty。

13）srvutil：服务器工具类。

14）store：消息存储实现相关类。

15）style：checkstyle 相关实现。

16）test：测试相关类。

17）tools：工具类，监控命令相关实现类。

1.3 RocketMQ 的设计理念和目标

1.3.1 设计理念

RocketMQ 设计基于主题的发布与订阅模式，其核心功能包括消息发送、消息存储（Broker）、消息消费，整体设计追求简单与性能第一，主要体现在如下三个方面。

首先，NameServer 设计极其简单，摒弃了业界常用的使用 Zookeeper 充当信息管理的"注册中心"，而是自研 NameServer 来实现元数据的管理（Topic 路由信息等）。从实际需求出发，因为 Topic 路由信息无须在集群之间保持强一致，追求最终一致性，并且能容忍分钟级的不一致。正是基于此种情况，RocketMQ 的 NameServer 集群之间互不通信，极大地降低了 NameServer 实现的复杂程度，对网络的要求也降低了不少，但是性能相比较 Zookeeper 有了极大的提升。

其次是高效的 IO 存储机制。RocketMQ 追求消息发送的高吞吐量，RocketMQ 的消息存储文件设计成文件组的概念，组内单个文件大小固定，方便引入内存映射机制，所有主题的消息存储基于顺序写，极大地提供了消息写性能，同时为了兼顾消息消费与消息查找，引入了消息消费队列文件与索引文件。

最后是容忍存在设计缺陷，适当将某些工作下放给 RocketMQ 使用者。消息中间件的实现者经常会遇到一个难题：如何保证消息一定能被消息消费者消费，并且保证只消费一次。RocketMQ 的设计者给出的解决办法是不解决这个难题，而是退而求其次，只保证消息被消费者消费，但设计上允许消息被重复消费，这样极大地简化了消息中间件的内核，使得实现消息发送高可用变得非常简单与高效，消息重复问题由消费者在消息消费时实现幂等。

1.3.2 设计目标

RocketMQ 作为一款消息中间件，需要解决如下问题。

1. 架构模式

RocketMQ 与大部分消息中间件一样，采用发布订阅模式，基本的参与组件主要包括：消息发送者、消息服务器（消息存储）、消息消费、路由发现。

2. 顺序消息

所谓顺序消息，就是消息消费者按照消息达到消息存储服务器的顺序消费。RocketMQ 可以严格保证消息有序。

3. 消息过滤

消息过滤是指在消息消费时，消息消费者可以对同一主题下的消息按照规则只消费自己感兴趣的消息。RocketMQ 消息过滤支持在服务端与消费端的消息过滤机制。

1）消息在 Broker 端过滤。Broker 只将消息消费者感兴趣的消息发送给消息消费者。

2）消息在消息消费端过滤，消息过滤方式完全由消息消费者自定义，但缺点是有很多无用的消息会从 Broker 传输到消费端。

4. 消息存储

消息中间件的一个核心实现是消息的存储，对消息存储一般有如下两个维度的考量：消息堆积能力和消息存储性能。RocketMQ 追求消息存储的高性能，引入内存映射机制，所有主题的消息顺序存储在同一个文件中。同时为了避免消息无限在消息存储服务器中累积，引入了消息文件过期机制与文件存储空间报警机制。

5. 消息高可用性

通常影响消息可靠性的有以下几种情况。

1）Broker 正常关机。

2）Broker 异常 Crash。

3）OS Crash。

4）机器断电，但是能立即恢复供电情况。

5）机器无法开机（可能是 CPU、主板、内存等关键设备损坏）。

6）磁盘设备损坏。

针对上述情况，情况 1~4 的 RocketMQ 在同步刷盘机制下可以确保不丢失消息，在异步刷盘模式下，会丢失少量消息。情况 5~6 属于单点故障，一旦发生，该节点上的消息全部丢失，如果开启了异步复制机制，RoketMQ 能保证只丢失少量消息，RocketMQ 在后续版本中将引入双写机制，以满足消息可靠性要求极高的场合。

6. 消息到达（消费）低延迟

RocketMQ 在消息不发生消息堆积时，以长轮询模式实现准实时的消息推送模式。

7. 确保消息必须被消费一次

RocketMQ 通过消息消费确认机制 (ACK) 来确保消息至少被消费一次，但由于 ACK 消息有可能丢失等其他原因，RocketMQ 无法做到消息只被消费一次，有重复消费的可能。

8. 回溯消息

回溯消息是指消息消费端已经消费成功的消息，由于业务要求需要重新消费消息。RocketMQ 支持按时间回溯消息，时间维度可精确到毫秒，可以向前或向后回溯。

9. 消息堆积

消息中间件的主要功能是异步解耦，必须具备应对前端的数据洪峰，提高后端系统的可用性，必然要求消息中间件具备一定的消息堆积能力。RocketMQ 消息存储使用磁盘文件（内存映射机制），并且在物理布局上为多个大小相等的文件组成逻辑文件组，可以无限循环使用。RocketMQ 消息存储文件并不是永久存储在消息服务器端，而是提供了过期机制，默认保留 3 天。

10. 定时消息

定时消息是指消息发送到 Broker 后，不能被消息消费端立即消费，要到特定的时间点或者等待特定的时间后才能被消费。如果要支持任意精度的定时消息消费，必须在消息服务端对消息进行排序，势必带来很大的性能损耗，故 RocketMQ 不支持任意进度的定时消息，而只支持特定延迟级别。

11. 消息重试机制

消息重试是指消息在消费时，如果发送异常，消息中间件需要支持消息重新投递，RocketMQ 支持消息重试机制。

第 2 章

RocketMQ 路由中心 NameServer

本章主要介绍 RocketMQ 路由管理、服务注册及服务发现的机制，NameServer 是整个 RocketMQ 的"大脑"。相信大家对"服务发现"这个词语并不陌生，分布式服务 SOA 架构体系中会有服务注册中心，分布式服务 SOA 的注册中心主要提供服务调用的解析服务，指引服务调用方（消费者）找到"远方"的服务提供者，完成网络通信，那么 RocketMQ 的路由中心存储的是什么数据呢？作为一款高性能的消息中间件，如何避免 NameServer 的单点故障，提供高可用性呢？让我们带着上述疑问，一起进入 RocketMQ NameServer 的精彩世界中来。

本章重点内容如下。
- NameServer 整体架构设计
- NameServer 动态路由发现与剔除机制

2.1 NameServer 架构设计

消息中间件的设计思路一般基于主题的订阅发布机制，消息生产者（Producer）发送某一主题的消息到消息服务器，消息服务器负责该消息的持久化存储，消息消费者（Consumer）订阅感兴趣的主题，消息服务器根据订阅信息（路由信息）将消息推送到消费者（PUSH 模式）或者消息消费者主动向消息服务器拉取消息（PULL 模式），从而实现消息生产者与消息消费者解耦。为了避免消息服务器的单点故障导致的整个系统瘫痪，通常会部署多台消息服务器共同承担消息的存储。那消息生产者如何知道消息要发往哪台消息服务器呢？如果某一台消息服务器宕机了，那么生产者如何在不重启服务的情况下感知呢？

NameServer 就是为了解决上述问题而设计的。

RocketMQ 的逻辑部署图如图 2-1 所示。

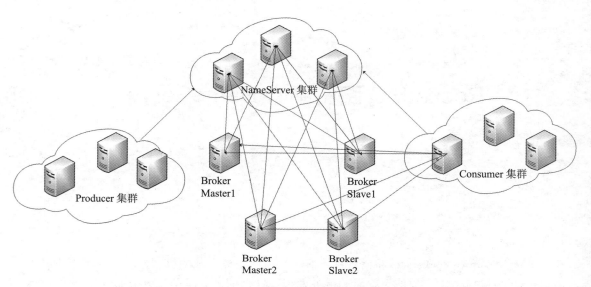

图 2-1 RocketMQ 物理部署图

Broker 消息服务器在启动时向所有 NameServer 注册，消息生产者 (Producer) 在发送消息之前先从 NameServer 获取 Broker 服务器地址列表，然后根据负载算法从列表中选择一台消息服务器进行消息发送。NameServer 与每台 Broker 服务器保持长连接，并间隔 30s 检测 Broker 是否存活，如果检测到 Broker 宕机，则从路由注册表中将其移除。但是路由变化不会马上通知消息生产者，为什么要这样设计呢？这是为了降低 NameServer 实现的复杂性，在消息发送端提供容错机制来保证消息发送的高可用性，这部分在 3.4 节中会有详细的描述。

NameServer 本身的高可用可通过部署多台 NameServer 服务器来实现，但彼此之间互不通信，也就是 NameServer 服务器之间在某一时刻的数据并不会完全相同，但这对消息发送不会造成任何影响，这也是 RocketMQ NameServer 设计的一个亮点，RocketMQ NameServer 设计追求简单高效。

2.2 NameServer 启动流程

从源码的角度窥探一下 NameServer 启动流程，重点关注 NameServer 相关启动参数。

NameServer 启动类：org.apache.rocketmq.namesrv.NamesrvStartup。

Step1：首先来解析配置文件，需要填充 NameServerConfig、NettyServerConfig 属性值。

代码清单 2-1　NameServer 加载配置文件

```
final NamesrvConfig namesrvConfig = new NamesrvConfig();
final NettyServerConfig nettyServerConfig = new NettyServerConfig();
nettyServerConfig.setListenPort(9876);
if (commandLine.hasOption('c')) {
    String file = commandLine.getOptionValue('c');
    if (file != null) {
        InputStream in = new BufferedInputStream(new FileInputStream(file));
        properties = new Properties();
        properties.load(in);
        MixAll.properties2Object(properties, namesrvConfig);
        MixAll.properties2Object(properties, nettyServerConfig);
        namesrvConfig.setConfigStorePath(file);
        System.out.printf("load config properties file OK, " + file + "%n");
        in.close();
    }
}
if (commandLine.hasOption('p')) {
    MixAll.printObjectProperties(null, namesrvConfig);
    MixAll.printObjectProperties(null, nettyServerConfig);
    System.exit(0);
}
MixAll.properties2Object(ServerUtil.commandLine2Properties(commandLine),
        namesrvConfig);
```

从代码我们可以知道先创建 NameServerConfig（NameServer 业务参数）、NettyServer-Config（NameServer 网络参数），然后在解析启动时把指定的配置文件或启动命令中的选项值，填充到 nameServerConfig、nettyServerConfig 对象。参数来源有如下两种方式。

1）-c configFile 通过 -c 命令指定配置文件的路径。

2）使用"-- 属性名 属性值"，例如 --listenPort 9876。

代码清单 2-2　NameServerConfig 属性

```
private String rocketmqHome = System.getProperty(MixAll.ROCKETMQ_HOME_PROPERTY,
        System.getenv(MixAll.ROCKETMQ_HOME_ENV));
private String kvConfigPath = System.getProperty("user.home") + File.separator
        + "namesrv" + File.separator + "kvConfig.json";
private String configStorePath = System.getProperty("user.home") +
    File.separator + "namesrv" + File.separator +
            "namesrv.properties";
private String productEnvName = "center";
private boolean clusterTest = false;
private boolean orderMessageEnable = false;
```

- rocketmqhome：rocketmq 主目录，可以通过 -Drocketmq.home.dir=path 或通过设置环境变量 ROCKETMQ_HOME 来配置 RocketMQ 的主目录。
- kvConfigPath：NameServer 存储 KV 配置属性的持久化路径。
- configStorePath:nameServer 默认配置文件路径，不生效。nameServer 启动时如果要通过配置文件配置 NameServer 启动属性的话，请使用 -c 选项。
- orderMessageEnable：是否支持顺序消息，默认是不支持。

代码清单 2-3　NettyServerConfig 属性

```
private int listenPort = 8888;
private int serverWorkerThreads = 8;
private int serverCallbackExecutorThreads = 0;
private int serverSelectorThreads = 3;
private int serverOnewaySemaphoreValue = 256;
private int serverAsyncSemaphoreValue = 64;
private int serverChannelMaxIdleTimeSeconds = 120;
private int serverSocketSndBufSize = NettySystemConfig.socketSndbufSize;
private int serverSocketRcvBufSize = NettySystemConfig.socketRcvbufSize;
private boolean serverPooledByteBufAllocatorEnable = true;
private boolean useEpollNativeSelector = false;
```

- listenPort：NameServer 监听端口，该值默认会被初始化为 9876。
- serverWorkerThreads：Netty 业务线程池线程个数。
- serverCallbackExecutorThreads：Netty public 任务线程池线程个数，Netty 网络设计，根据业务类型会创建不同的线程池，比如处理消息发送、消息消费、心跳检测等。如果该业务类型（RequestCode）未注册线程池，则由 public 线程池执行。
- serverSelectorThreads：IO 线程池线程个数，主要是 NameServer、Broker 端解析请求、返回相应的线程个数，这类线程主要是处理网络请求的，解析请求包，然后转发到各个业务线程池完成具体的业务操作，然后将结果再返回调用方。
- serverOnewaySemaphoreValue：send oneway 消息请求并发度（Broker 端参数）。
- serverAsyncSemaphoreValue：异步消息发送最大并发度（Broker 端参数）。
- serverChannelMaxIdleTimeSeconds：网络连接最大空闲时间，默认 120s。如果连接空闲时间超过该参数设置的值，连接将被关闭。
- serverSocketSndBufSize：网络 socket 发送缓存区大小，默认 64k。
- serverSocketRcvBufSize：网络 socket 接收缓存区大小，默认 64k。
- serverPooledByteBufAllocatorEnable：ByteBuffer 是否开启缓存，建议开启。
- useEpollNativeSelector：是否启用 Epoll IO 模型，Linux 环境建议开启。

在启动 NameServer 时，可以先使用 ./mqnameserver -c configFile -p 打印当前加载的配置属性。

Step2：根据启动属性创建 NamesrvController 实例，并初始化该实例，NameServerController 实例为 NameServer 核心控制器。

代码清单 2-4　NamesrvController#Initialize 代码片段

```
public boolean initialize() {
    this.kvConfigManager.load();
    this.remotingServer = new NettyRemotingServer(this.nettyServerConfig,
            this.brokerHousekeepingService);
    this.remotingExecutor =
        Executors.newFixedThreadPool(nettyServerConfig.getServerWorkerThreads(),
        new ThreadFactoryImpl("RemotingExecutorThread_"));
    this.registerProcessor();
    this.scheduledExecutorService.scheduleAtFixedRate(new Runnable() {
            public void run() {
                NamesrvController.this.routeInfoManager.scanNotActiveBroker();
            }
    }, 5, 10, TimeUnit.SECONDS);

    this.scheduledExecutorService.scheduleAtFixedRate(new Runnable() {
            public void run() {
                NamesrvController.this.kvConfigManager.printAllPeriodically();
            }
    }, 1, 10, TimeUnit.MINUTES);

    return true;
}
```

加载 KV 配置，创建 NettyServer 网络处理对象，然后开启两个定时任务，在 RocketMQ 中此类定时任务统称为心跳检测。

❑ 定时任务 1：NameServer 每隔 10s 扫描一次 Broker，移除处于不激活状态的 Broker。
❑ 定时任务 2：nameServer 每隔 10 分钟打印一次 KV 配置。

Step3：注册 JVM 钩子函数并启动服务器，以便监听 Broker、消息生产者的网络请求。

代码清单 2-5　注册 JVM 钩子函数代码

```
Runtime.getRuntime().addShutdownHook(new ShutdownHookThread(log, new
    Callable<Void>() {
        public Void call() throws Exception {
            controller.shutdown();
            return null;
        }
}));
controller.start();
```

这里主要是向读者展示一种常用的编程技巧，如果代码中使用了线程池，一种优雅停机的方式就是注册一个 JVM 钩子函数，在 JVM 进程关闭之前，先将线程池关闭，及时释放资源。

2.3 NameServer 路由注册、故障剔除

NameServer 主要作用是为消息生产者和消息消费者提供关于主题 Topic 的路由信息，那么 NameServer 需要存储路由的基础信息，还要能够管理 Broker 节点，包括路由注册、路由删除等功能。

2.3.1 路由元信息

NameServer 路由实现类：org.apache.rocketmq.namesrv.routeinfo.RouteInfoManager，在了解路由注册之前，我们首先看一下 NameServer 到底存储哪些信息。

代码清单 2-6　RouteInfoManager 路由元数据

```
private final HashMap<String/* topic */, List<QueueData>> topicQueueTable;
private final HashMap<String/* brokerName */, BrokerData> brokerAddrTable;
private final HashMap<String/* clusterName */, Set<String/* brokerName */>>
    clusterAddrTable;
private final HashMap<String/* brokerAddr */, BrokerLiveInfo> brokerLiveTable;
private final HashMap<String/* brokerAddr */, List<String>/* Filter Server */>
    filterServerTable;
```

- topicQueueTable：Topic 消息队列路由信息，消息发送时根据路由表进行负载均衡。
- brokerAddrTable：Broker 基础信息，包含 brokerName、所属集群名称、主备 Broker 地址。
- clusterAddrTable：Broker 集群信息，存储集群中所有 Broker 名称。
- brokerLiveTable：Broker 状态信息。NameServer 每次收到心跳包时会替换该信息。
- filterServerTable：Broker 上的 FilterServer 列表，用于类模式消息过滤，详细介绍请参考第 6 章的内容。

QueueData、BrokerData、BrokerLiveInfo 类图如图 2-2 所示。

RocketMQ 基于订阅发布机制，一个 Topic 拥有多个消息队列，一个 Broker 为每一主题默认创建 4 个读队列 4 个写队列。多个 Broker 组成一个集群，BrokerName 由相同的多台 Broker 组成 Master-Slave 架构，brokerId 为 0 代表 Master，大于 0 表示 Slave。BrokerLiveInfo 中的 lastUpdateTimestamp 存储上次收到 Broker 心跳包的时间。

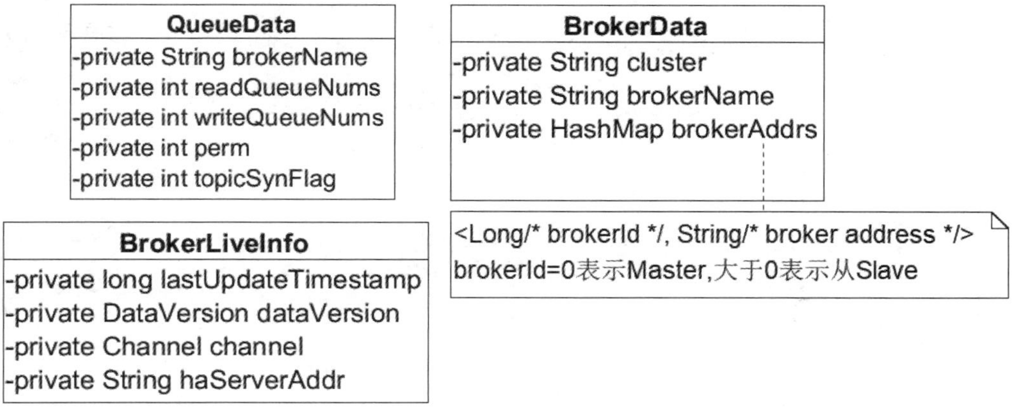

图 2-2　路由元数据类图

RocketMQ2 主 2 从部署图如图 2-3 所示。

图 2-3　RocketMQ　2 主 2 从数据结构展示图

对应运行时数据结构如图 2-4 和图 2-5 所示。

```
topicQueueTable:{
    "topic1":[
        {
            "brokerName":"broker-a",
            "readQueueNums":4,
            "readQueueNums":4,
            "perm":6,    // 读写权限，具体含义请参考PermName
            "topicSynFlag":0 // topic同步标记，具体含义请参考TopicSysFlag
        },
        {
            "brokerName":"broker-b",
            "readQueueNums":4,
            "readQueueNums":4,
            "perm":6,    // 读写权限，具体含义请参考PermName
            "topicSynFlag":0 // topic同步标记，具体含义请参考TopicSysFlag
        }
    ],
    "topic other":[]
}

brokerAddrTable:{
    "broker-a": {
        "cluster":"c1",
        "brokerName":"broker-a",
        "brokerAddrs":{
            0:"192.168.56.1:10000",
            1:"192.168.56.2:10000",
        }
    },
    "broker-b": {
        "cluster":"c1",
        "brokerAddrs":{
            0:"192.168.56.3:10000",
            1:"192.168.56.4:10000"
        }
    }
}
```

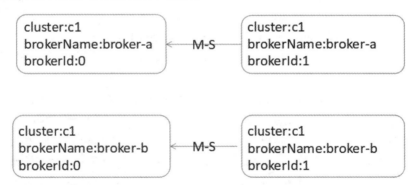

图 2-4　TopicQueueTable、brokerAddrTable 运行时内存结构

```
brokerLiveTable : {                                    clusterAddrTable: {
    "192.168.56.1:10000" : {                               "c1" : [{"broker-a", "broker-b"}]
        "lastUpdateTimestamp":1518270318980,           }
        "dataVersion":versionObl,
        "channel":channelObj,
        "haServerAddr":"192.168.56.2:10000"
    },
    "192.168.56.2:10000" : {
        "lastUpdateTimestamp":1518270318980,
        "dataVersion":versionObl,
        "channel":channelObj,
        "haServerAddr":""
    },
    "192.168.56.3:10000" : {
        "lastUpdateTimestamp":1518270318980,
        "dataVersion":versionObl,
        "channel":channelObj,
        "haServerAddr":"192.168.56.4:10000"
    },
    "192.168.56.4:10000" : {
        "lastUpdateTimestamp":1518270318980,
        "dataVersion":versionObl,
        "channel":channelObj,
        "haServerAddr":""
    },
}
```

图 2-5　BrokerLiveTable、clusterAddrTable 运行时内存结构

2.3.2　路由注册

RocketMQ 路由注册是通过 Broker 与 NameServer 的心跳功能实现的。Broker 启动时向集群中所有的 NameServer 发送心跳语句，每隔 30s 向集群中所有 NameServer 发送心跳包，NameServer 收到 Broker 心跳包时会更新 brokerLiveTable 缓存中 BrokerLiveInfo 的 lastUpdateTimestamp，然后 Name Server 每隔 10s 扫描 brokerLiveTable，如果连续 120s 没有收到心跳包，NameServer 将移除该 Broker 的路由信息同时关闭 Socket 连接。

1. Broker 发送心跳包

Broker 发送心跳包的核心代码如下所示。

代码清单 2-7　Broker 端心跳包发送（BrokerController#start）

```
this.scheduledExecutorService.scheduleAtFixedRate(new Runnable() {
        public void run() {
            try {
                BrokerController.this.registerBrokerAll(true, false);
            } catch (Throwable e) {
                log.error("registerBrokerAll Exception", e);
            }
        }
}, 1000 * 10, 1000 * 30, TimeUnit.MILLISECONDS);
```

代码清单 2-8　BrokerOuterAPI#registerBrokerAll

```
List<String> nameServerAddressList =
    this.remotingClient.getNameServerAddressList();
```

```
if (nameServerAddressList != null) {
    for (String namesrvAddr : nameServerAddressList) {// 遍历所有 NameServer 列表
        try {
            RegisterBrokerResult result = this.registerBroker(namesrvAddr,
                clusterName, brokerAddr, brokerName, brokerId,
                haServerAddr, topicConfigWrapper, filterServerList, oneway,
                timeoutMills); // 分别向 NameServer 注册
            if (result != null) {
                registerBrokerResult = result;
            }
            log.info("register broker to name server {} OK", namesrvAddr);
        } catch (Exception e) {
            log.warn("registerBroker Exception, {}", namesrvAddr, e);
        }
    }
}
```

该方法主要是遍历 NameServer 列表，Broker 消息服务器依次向 NameServer 发送心跳包。

代码清单 2-9　BrokerOuterAPI#registerBroker（网络发送代码）

```
RegisterBrokerRequestHeader requestHeader = new RegisterBrokerRequestHeader();
requestHeader.setBrokerAddr(brokerAddr);
requestHeader.setBrokerId(brokerId);
requestHeader.setBrokerName(brokerName);
requestHeader.setClusterName(clusterName);
requestHeader.setHaServerAddr(haServerAddr);
RemotingCommand request = RemotingCommand.createRequestCommand(
        RequestCode.REGISTER_BROKER, requestHeader);
RegisterBrokerBody requestBody = new RegisterBrokerBody();
requestBody.setTopicConfigSerializeWrapper(topicConfigWrapper);
requestBody.setFilterServerList(filterServerList);
request.setBody(requestBody.encode());
if (oneway) {
    try {
            this.remotingClient.invokeOneway(namesrvAddr, request, timeoutMills);
    } catch (RemotingTooMuchRequestException e) {
            // Ignore
    }
    return null;
}
RemotingCommand response = this.remotingClient.invokeSync(namesrvAddr, request,
        timeoutMills);
```

发送心跳包具体逻辑，首先封装请求包头 (Header)。

- brokerAddr：broker 地址。
- brokerId：brokerId,0:Master；大于 0：Slave。
- brokerName：broker 名称。
- clusterName：集群名称。

❑ haServerAddr：master 地址，初次请求时该值为空，slave 向 Nameserver 注册后返回。
❑ requestBody：
- filterServerList。消息过滤服务器列表。
- topicConfigWrapper。主题配置，topicConfigWrapper 内部封装的是 TopicConfig-Manager 中的 topicConfigTable，内部存储的是 Broker 启动时默认的一些 Topic，MixAll.SELF_TEST_TOPIC、MixAll.DEFAULT_TOPIC（AutoCreateTopic-Enable=true）、MixAll.BENCHMARK_TOPIC、MixAll.OFFSET_MOVED_EVENT、BrokerConfig#brokerClusterName、BrokerConfig#brokerName。Broker 中 Topic 默认存储在 ${Rocket_Home}/store/confg/topic.json 中。

RocketMQ 网络传输基于 Netty，具体网络实现细节本书不会过细去剖析，在这里介绍一下网络跟踪方法：每一个请求，RocketMQ 都会定义一个 RequestCode，然后在服务端会对应相应的网络处理器 (processor 包中)，只需整库搜索 RequestCode 即可找到相应的处理逻辑。如果对 Netty 感兴趣，可以参考笔者发布的《源码研究 Netty 系列》（http://blog.csdn.net/column/details/15042.html）。

2. NameServer 处理心跳包

org.apache.rocketmq.namesrv.processor.DefaultRequestProcessor 网络处理器解析请求类型，如果请求类型为 RequestCode.REGISTER_BROKER，则请求最终转发到 RouteInfoManager#registerBroker。

代码清单 2-10　RouteInfoManager#registerBroker clusterAddrTable 维护

```
this.lock.writeLock().lockInterruptibly();
Set<String> brokerNames = this.clusterAddrTable.get(clusterName);
if (null == brokerNames) {
    brokerNames = new HashSet<String>();
    this.clusterAddrTable.put(clusterName, brokerNames);
}
brokerNames.add(brokerName);
```

Step1：路由注册需要加写锁，防止并发修改 RouteInfoManager 中的路由表。首先判断 Broker 所属集群是否存在，如果不存在，则创建，然后将 broker 名加入到集群 Broker 集合中。

代码清单 2-11　RouteInfoManager#registerBroker brokerAddrTable 维护

```
BrokerData brokerData = this.brokerAddrTable.get(brokerName);
if (null == brokerData) {
        registerFirst = true;
        brokerData = new BrokerData(clusterName, brokerName, new HashMap<Long,
            String>());
        this.brokerAddrTable.put(brokerName, brokerData);
    }
```

```
        String oldAddr = brokerData.getBrokerAddrs().put(brokerId, brokerAddr);
        registerFirst = registerFirst || (null == oldAddr);
```

Step2：维护 BrokerData 信息，首先从 brokerAddrTable 根据 BrokerName 尝试获取 Broker 信息，如果不存在，则新建 BrokerData 并放入到 brokerAddrTable，registerFirst 设置为 true；如果存在，直接替换原先的，registerFirst 设置为 false，表示非第一次注册。

代码清单 2-12　RouteInfoManager#registerBroker topicQueueTable 维护

```
if (null != topicConfigWrapper && MixAll.MASTER_ID == brokerId) {
    if (this.isBrokerTopicConfigChanged(brokerAddr,
            topicConfigWrapper.getDataVersion()) || registerFirst) {
        ConcurrentMap<String, TopicConfig> tcTable =
                topicConfigWrapper.getTopicConfigTable();
        if (tcTable != null) {
            for (Map.Entry<String, TopicConfig> entry : tcTable.entrySet()) {
                this.createAndUpdateQueueData(brokerName, entry.getValue());
            }
        }
    }
}
```

Step3：如果 Broker 为 Master，并且 Broker Topic 配置信息发生变化或者是初次注册，则需要创建或更新 Topic 路由元数据，填充 topicQueueTable，其实就是为默认主题自动注册路由信息，其中包含 MixAll.DEFAULT_TOPIC 的路由信息。当消息生产者发送主题时，如果该主题未创建并且 BrokerConfig 的 autoCreateTopicEnable 为 true 时，将返回 MixAll.DEFAULT_TOPIC 的路由信息。

代码清单 2-13　RouteInfoManager#createAndUpdateQueueData

```
private void createAndUpdateQueueData(final String brokerName, final TopicConfig
topicConfig) {
    QueueData queueData = new QueueData();
    queueData.setBrokerName(brokerName);
    queueData.setWriteQueueNums(topicConfig.getWriteQueueNums());
    queueData.setReadQueueNums(topicConfig.getReadQueueNums());
    queueData.setPerm(topicConfig.getPerm());
    queueData.setTopicSynFlag(topicConfig.getTopicSysFlag());
    List<QueueData> queueDataList =
                this.topicQueueTable.get(topicConfig.getTopicName());
    if (null == queueDataList) {
        queueDataList = new LinkedList<QueueData>();
        queueDataList.add(queueData);
        this.topicQueueTable.put(topicConfig.getTopicName(),
            queueDataList);
        log.info("new topic registerd, {} {}", topicConfig.getTopicName(),
            queueData);
    } else {
        boolean addNewOne = true;
```

```
            Iterator<QueueData> it = queueDataList.iterator();
            while (it.hasNext()) {
                QueueData qd = it.next();
                if (qd.getBrokerName().equals(brokerName)) {
                    if (qd.equals(queueData)) {
                        addNewOne = false;
                    } else {
                        log.info("topic changed, {} OLD: {} NEW: {}",
                            topicConfig.getTopicName(), qd, queueData);
                        it.remove();
                    }
                }
            }

            if (addNewOne) {
                queueDataList.add(queueData);
            }
        }
    }
```

根据 TopicConfig 创建 QueueData 数据结构，然后更新 topicQueueTable。

代码清单 2-14　RouteInfoManager#registerBroker

```
BrokerLiveInfo prevBrokerLiveInfo = this.brokerLiveTable.put(brokerAddr,
    new BrokerLiveInfo(System.currentTimeMillis(),
        topicConfigWrapper.getDataVersion(),
        channel,
        haServerAddr));
if (null == prevBrokerLiveInfo) {
    log.info("new broker registerd, {} HAServer: {}", brokerAddr, haServerAddr);
}
```

Step4：更新 BrokerLiveInfo，存活 Broker 信息表，BrokeLiveInfo 是执行路由删除的重要依据。

代码清单 2-15　RouteInfoManager#registerBroker

```
if (filterServerList != null) {
    if (filterServerList.isEmpty()) {
            this.filterServerTable.remove(brokerAddr);
    } else {
            this.filterServerTable.put(brokerAddr, filterServerList);
    }
}
if (MixAll.MASTER_ID != brokerId) {
    String masterAddr = brokerData.getBrokerAddrs().get(MixAll.MASTER_ID);
        if (masterAddr != null) {
        BrokerLiveInfo brokerLiveInfo = this.brokerLiveTable.get(masterAddr);
        if (brokerLiveInfo != null) {
            result.setHaServerAddr(brokerLiveInfo.getHaServerAddr());
```

```
                result.setMasterAddr(masterAddr);
            }
        }
    }
```

Step5：注册 Broker 的过滤器 Server 地址列表，一个 Broker 上会关联多个 FilterServer 消息过滤服务器，此部分内容将在第 6 章详细介绍；如果此 Broker 为从节点，则需要查找该 Broker 的 Master 的节点信息，并更新对应的 masterAddr 属性。

设计亮点：NameServe 与 Broker 保持长连接，Broker 状态存储在 brokerLiveTable 中，NameServer 每收到一个心跳包，将更新 brokerLiveTable 中关于 Broker 的状态信息以及路由表（topicQueueTable、brokerAddrTable、brokerLiveTable、filterServerTable）。更新上述路由表（HashTable）使用了锁粒度较少的读写锁，允许多个消息发送者（Producer）并发读，保证消息发送时的高并发。但同一时刻 NameServer 只处理一个 Broker 心跳包，多个心跳包请求串行执行。这也是读写锁经典使用场景，更多关于读写锁的信息，可以参考笔者的博文：http://blog.csdn.net/prestigeding/article/details/53286756。

2.3.3 路由删除

根据上面章节的介绍，Broker 每隔 30s 向 NameServer 发送一个心跳包，心跳包中包含 BrokerId、Broker 地址、Broker 名称、Broker 所属集群名称、Broker 关联的 FilterServer 列表。但是如果 Broker 宕机，NameServer 无法收到心跳包，此时 NameServer 如何来剔除这些失效的 Broker 呢？Name Server 会每隔 10s 扫描 brokerLiveTable 状态表，如果 BrokerLive 的 lastUpdateTimestamp 的时间戳距当前时间超过 120s，则认为 Broker 失效，移除该 Broker，关闭与 Broker 连接，并同时更新 topicQueueTable、brokerAddrTable、brokerLiveTable、filterServerTable。

RocktMQ 有两个触发点来触发路由删除。

1）NameServer 定时扫描 brokerLiveTable 检测上次心跳包与当前系统时间的时间差，如果时间戳大于 120s，则需要移除该 Broker 信息。

2）Broker 在正常被关闭的情况下，会执行 unregisterBroker 指令。

由于不管是何种方式触发的路由删除，路由删除的方法都是一样的，就是从 topicQueueTable、brokerAddrTable、brokerLiveTable、filterServerTable 删除与该 Broker 相关的信息，但 RocketMQ 这两种方式维护路由信息时会抽取公共代码，本文将以第一种方式展开分析。

代码清单 2-16　RouteInfoManager#scanNotActiveBroker

```
public void scanNotActiveBroker() {
    Iterator<Entry<String, BrokerLiveInfo>> it =
```

```
            this.brokerLiveTable.entrySet().iterator();
while (it.hasNext()) {
    Entry<String, BrokerLiveInfo> next = it.next();
    long last = next.getValue().getLastUpdateTimestamp();
    if ((last + BROKER_CHANNEL_EXPIRED_TIME) < System.currentTimeMillis()) {
        RemotingUtil.closeChannel(next.getValue().getChannel());
        it.remove();
        log.warn("The broker channel expired, {} {}ms", next.getKey(),
            BROKER_CHANNEL_EXPIRED_TIME);
        this.onChannelDestroy(next.getKey(), next.getValue().getChannel());
    }
}
```

我们应该不会忘记 scanNotActiveBroker 在 NameServer 中每 10s 执行一次。逻辑也很简单，遍历 brokerLiveInfo 路由表（HashMap），检测 BrokerLiveInfo 的 lastUpdateTimestamp 上次收到心跳包的时间如果超过当前时间 120s，NameServer 则认为该 Broker 已不可用，故需要将它移除，关闭 Channel，然后删除与该 Broker 相关的路由信息，路由表维护过程，需要申请写锁。

代码清单 2-17 RouteInfoManager#onChannelDestroy

```
this.lock.writeLock().lockInterruptibly();
this.brokerLiveTable.remove(brokerAddrFound);
this.filterServerTable.remove(brokerAddrFound);
```

Step1：申请写锁，根据 brokerAddress 从 brokerLiveTable、filterServerTable 移除，如代码清单 2-18 所示。

代码清单 2-18 RouteInfoManager#onChannelDestroy

```
String brokerNameFound = null;
boolean removeBrokerName = false;
Iterator<Entry<String, BrokerData>> itBrokerAddrTable =
    this.brokerAddrTable.entrySet().iterator();
while (itBrokerAddrTable.hasNext() && (null == brokerNameFound)) {
    BrokerData brokerData = itBrokerAddrTable.next().getValue();
    Iterator<Entry<Long, String>> it =
        brokerData.getBrokerAddrs().entrySet().iterator();
    while (it.hasNext()) {
        Entry<Long, String> entry = it.next();
        Long brokerId = entry.getKey();
        String brokerAddr = entry.getValue();
        if (brokerAddr.equals(brokerAddrFound)) {
            brokerNameFound = brokerData.getBrokerName();
            it.remove();
            log.info("remove brokerAddr[{}, {}] from brokerAddrTable, 
                because channel destroyed",
                brokerId, brokerAddr);
```

```
            break;
        }
    }
    if (brokerData.getBrokerAddrs().isEmpty()) {
        removeBrokerName = true;
        itBrokerAddrTable.remove();
        log.info("remove brokerName[{}] from brokerAddrTable, because channel
            destroyed",brokerData.getBrokerName());
    }
}
```

Step2：维护 brokerAddrTable。遍历从 HashMap<String/* brokerName */, BrokerData> brokerAddrTable，从 BrokerData 的 HashMap<Long/* brokerId */, String/* broker address */> brokerAddrs 中，找到具体的 Broker，从 BrokerData 中移除，如果移除后在 BrokerData 中不再包含其他 Broker，则在 brokerAddrTable 中移除该 brokerName 对应的条目。

代码清单 2-19　RouteInfoManager#onChannelDestroy

```
if (brokerNameFound != null && removeBrokerName) {
    Iterator<Entry<String, Set<String>>> it =
            this.clusterAddrTable.entrySet().iterator();
    while (it.hasNext()) {
        Entry<String, Set<String>> entry = it.next();
        String clusterName = entry.getKey();
        Set<String> brokerNames = entry.getValue();
        boolean removed = brokerNames.remove(brokerNameFound);
        if (removed) {
            log.info("remove brokerName[{}], clusterName[{}] from
                clusterAddrTable, because channel destroyed",
                    brokerNameFound, clusterName);

            if (brokerNames.isEmpty()) {
                log.info("remove the clusterName[{}] from clusterAddrTable,
                    because channel destroyed and no broker in this cluster",
                        clusterName);
                it.remove();
            }
            break;
        }
    }
}
```

Step3：根据 BrokerName，从 clusterAddrTable 中找到 Broker 并从集群中移除。如果移除后，集群中不包含任何 Broker，则将该集群从 clusterAddrTable 中移除。

代码清单 2-20　RouteInfoManager#onChannelDestroy

```
if (removeBrokerName) {
    Iterator<Entry<String, List<QueueData>>> itTopicQueueTable =
            this.topicQueueTable.entrySet().iterator();
    while (itTopicQueueTable.hasNext()) {
```

```
            Entry<String, List<QueueData>> entry = itTopicQueueTable.next();
            String topic = entry.getKey();
            List<QueueData> queueDataList = entry.getValue();
            Iterator<QueueData> itQueueData = queueDataList.iterator();
            while (itQueueData.hasNext()) {
                QueueData queueData = itQueueData.next();
                if (queueData.getBrokerName().equals(brokerNameFound)) {
                    itQueueData.remove();
                    log.info("remove topic[{} {}], from topicQueueTable, because
                        channel destroyed",topic, queueData);
                }
            }

            if (queueDataList.isEmpty()) {
                itTopicQueueTable.remove();
                log.info("remove topic[{}] all queue, from topicQueueTable, because
                    channel destroyed",topic);
            }
        }
    }
}
```

Step4：根据 brokerName，遍历所有主题的队列，如果队列中包含了当前 Broker 的队列，则移除，如果 topic 只包含待移除 Broker 的队列的话，从路由表中删除该 topic，如代码清单 2-21 所示。

代码清单 2-21　RouteInfoManager#onChannelDestroy

```
finally {
    this.lock.writeLock().unlock();
}
```

Step5：释放锁，完成路由删除。

2.3.4　路由发现

RocketMQ 路由发现是非实时的，当 Topic 路由出现变化后，NameServer 不主动推送给客户端，而是由客户端定时拉取主题最新的路由。根据主题名称拉取路由信息的命令编码为：GET_ROUTEINTO_BY_TOPIC。RocketMQ 路由结果如图 2-6 所示。

TopicRouteData
-private String orderTopicConf
-private List queueDatas
-private List brokerDatas
-private HashMap filterServerTable

图 2-6　RocketMQ 路由结果实体

❏ orderTopicConf：顺序消息配置内容，来自于 kvConfig。

❏ List<QueueData> queueData：topic 队列元数据。

❏ List<BrokerData> brokerDatas：topic 分布的 broker 元数据。

❏ HashMap<String/* brokerAdress*/,List<String> /*filterServer*/>：broker 上过滤服务

器地址列表。
- NameServer 路由发现实现类：DefaultRequestProcessor#getRouteInfoByTopic，如代码清单 2-22 所示。

代码清单 2-22　DefaultRequestProcessor#getRouteInfoByTopic

```java
public RemotingCommand getRouteInfoByTopic(ChannelHandlerContext ctx,
        RemotingCommand request) throws RemotingCommandException {
    final RemotingCommand response = RemotingCommand.createResponseCommand(null);
    final GetRouteInfoRequestHeader requestHeader =(GetRouteInfoRequestHeader)
        request.decodeCommandCustomHeader(GetRouteInfoRequestHeader.class);
    TopicRouteData topicRouteData = this.namesrvController.
        getRouteInfoManager().pickupTopicRouteData(requestHeader.getTopic());
    if (topicRouteData != null) {
        if(this.namesrvController.getNamesrvConfig().isOrderMessageEnable()) {
            String orderTopicConf =this.namesrvController.getKvConfigManager()
                .getKVConfig(NamesrvUtil.NAMESPACE_ORDER_TOPIC_CONFIG,
                requestHeader.getTopic());
            topicRouteData.setOrderTopicConf(orderTopicConf);
        }
        byte[] content = topicRouteData.encode();
        response.setBody(content);
        response.setCode(ResponseCode.SUCCESS);
        response.setRemark(null);
        return response;
    }
    response.setCode(ResponseCode.TOPIC_NOT_EXIST);
    response.setRemark("No topic route info in name server for the topic: "
            + requestHeader.getTopic()
            + FAQUrl.suggestTodo(FAQUrl.APPLY_TOPIC_URL));
    return response;
}
```

Step1：调用 RouterInfoManager 的方法，从路由表 topicQueueTable、brokerAddrTable、filterServerTable 中分别填充 TopicRouteData 中的 List<QueueData>、List<BrokerData> 和 filterServer 地址表。

Step2：如果找到主题对应的路由信息并且该主题为顺序消息，则从 NameServer KVconfig 中获取关于顺序消息相关的配置填充路由信息。

如果找不到路由信息 CODE 则使用 TOPIC_NOT_EXISTS，表示没有找到对应的路由。

2.4　本章小结

本章主要介绍了 NameServer 路由功能，包含路由元数据、路由注册与发现机制。为了加强对本章的理解，路由发现机制可以用图 2-6 来形象解释。

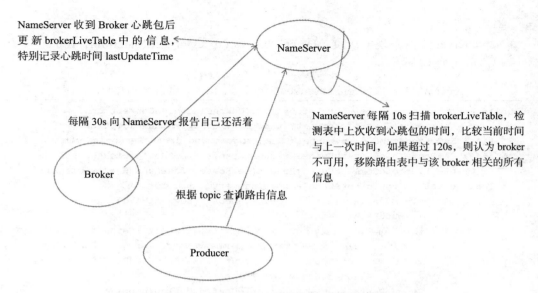

图 2-7　NameServer 路由注册、删除机制

NameServer 路由发现与删除机制就介绍到这里了，我们会发现这种设计会存在这样一种情况：NameServer 需要等 Broker 失效至少 120s 才能将该 Broker 从路由表中移除掉，那如果在 Broker 故障期间，消息生产者 Producer 根据主题获取到的路由信息包含已经宕机的 Broker，会导致消息发送失败，那这种情况怎么办，岂不是消息发送不是高可用的？让我们带着这个疑问进入 RocketMQ 消息发送的学习。

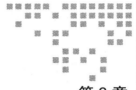

第 3 章 RocketMQ 消息发送

RocketMQ 发送普通消息有三种实现方式：可靠同步发送、可靠异步发送、单向 (Oneway) 发送。第 3 章主要聚焦在 RocketMQ 如何发送消息，然后从消息的数据结构开始，逐步介绍生产者的启动流程和消息发送的流程，最后再详细阐述批量消息发送。

本章重点内容如下。

- RocketMQ 消息结构
- 消息生产者 (Producer) 启动流程
- 消息发送过程
- 批量消息发送

3.1 漫谈 RocketMQ 消息发送

RocketMQ 支持 3 种消息发送方式：同步 (sync)、异步 (async)、单向 (oneway)。

同步：发送者向 MQ 执行发送消息 API 时，同步等待，直到消息服务器返回发送结果。

异步：发送者向 MQ 执行发送消息 API 时，指定消息发送成功后的回掉函数，然后调用消息发送 API 后，立即返回，消息发送者线程不阻塞，直到运行结束，消息发送成功或失败的回调任务在一个新的线程中执行。

单向：消息发送者向 MQ 执行发送消息 API 时，直接返回，不等待消息服务器的结果，也不注册回调函数，简单地说，就是只管发，不在乎消息是否成功存储在消息服务器上。

RocketMQ 消息发送需要考虑以下几个问题。

❑ 消息队列如何进行负载？

- 消息发送如何实现高可用？
- 批量消息发送如何实现一致性？

3.2 认识 RocketMQ 消息

RocketMQ 消息封装类是 org.apache.rocketmq.common.message.Message。
Message 类设计如图 3-1 所示。

```
Message
-private String topic
-private int flag
-private Map properties
-private byte[]
+public Message(String topic, byte [])
+public Message(String topic, String tags, String keys, int flag, byte [], boolean waitStoreMsgOK)
+public Message(String topic, String tags, byte [])
+public Message(String topic, String tags, String keys, byte [])
+public void setKeys(String keys)
+public void putUserProperty(final String name, final String value)
+public String getUserProperty(final String name)
+public int getDelayTimeLevel()
+public void setDelayTimeLevel(int level)
```

图 3-1 RocketMQ 消息类图

Message 的基础属性主要包括消息所属主题 topic、消息 Flag(RocketMQ 不做处理)、扩展属性、消息体。

RocketMQ 定义的 MessageFlag 如图 3-2 所示。

```
MessageSysFlag
-public final static int COMPRESSED_FLAG = 0x1
-public final static int MULTI_TAGS_FLAG = 0x1 << 1
-public final static int TRANSACTION_PREPARED_TYPE = 0x1 << 2
-public final static int TRANSACTION_COMMIT_TYPE = 0x2 << 2
-public final static int TRANSACTION_ROLLBACK_TYPE = 0x3 << 2
-public final static int TRANSACTION_NOT_TYPE = 0
+public static int getTransactionValue(final int flag)
+public static int resetTransactionValue(final int flag, final int type)
+public static int clearCompressedFlag(final int flag)
```

图 3-2 RocketMQ MessageSysFlag

代码清单 3-1 Message 全属性构造函数：

```
public Message(String topic, String tags, String keys, int flag, byte[] body,
    boolean waitStoreMsgOK) {
    this.topic = topic;
    this.flag = flag;
```

```java
        this.body = body;
        if (tags != null && tags.length() > 0)
            this.setTags(tags);
        if (keys != null && keys.length() > 0)
              this.setKeys(keys);

        this.setWaitStoreMsgOK(waitStoreMsgOK);
}
public void setTags(String tags) {
    this.putProperty(MessageConst.PROPERTY_TAGS, tags);
}
public void setKeys(Collection<String> keys) {
    StringBuffer sb = new StringBuffer();
    for (String k : keys) {
        sb.append(k);
        sb.append(MessageConst.KEY_SEPARATOR);
    }
    this.setKeys(sb.toString().trim());
}
```

Message 扩展属性主要包含下面几个。

tag：消息 TAG，用于消息过滤。

keys：Message 索引键，多个用空格隔开，RocketMQ 可以根据这些 key 快速检索到消息。

waitStoreMsgOK：消息发送时是否等消息存储完成后再返回。

delayTimeLevel：消息延迟级别，用于定时消息或消息重试。

这些扩展属性存储在 Message 的 properties 中。

3.3 生产者启动流程

消息生产者的代码都在 client 模块中，相对于 RocketMQ 来说，它就是客户端，也是消息的提供者，我们在应用系统中初始化生产者的一个实例即可使用它来发消息。

3.3.1 初识 DefaultMQProducer 消息发送者

DefaultMQProducer 是默认的消息生产者实现类，它实现 MQAdmin 的接口，其主要接口一览如图 3-3 和图 3-4 所示。

下面介绍 DefaultMQProducer 的主要方法。

1）void createTopic(String key, String newTopic, int queueNum, int topicSysFlag)
创建主题。

key：目前未实际作用，可以与 newTopic 相同。

newTopic：主题名称。

queueNum：队列数量。

topicSysFlag：主题系统标签，默认为 0。

```
<<Interface>>
MQAdmin
+void createTopic(String key, String newTopic, int queueNum)
+void createTopic(String key, String newTopic, int queueNum, int topicSysFlag)
+long searchOffset(final MessageQueue mq, final long timestamp)
+long maxOffset(final MessageQueue mq)
+long minOffset(final MessageQueue mq)
+long earliestMsgStoreTime(final MessageQueue mq)
+MessageExt viewMessage(final String offsetMsgId)
+QueryResult queryMessage(String topic, String key, int maxNum, long begin, long end)
+MessageExt viewMessage(String topic, String msgId)
```

图 3-3　MQAdmin

```
<<Interface>>
MQProducer
+List fetchPublishMessageQueues(final String topic)
+SendResult send(Message msg)
+SendResult send(Message msg, final long timeout)
+void send(Message msg, SendCallback sendCallback)
+void send(Message msg, SendCallback sendCallback, long timeout)
+void sendOneway(Message msg)
+SendResult send(Message msg, MessageQueue mq, final long timeout)
+void sendOneway(final Message msg, final MessageQueue mq)
+void send(final Message msg, final MessageQueue mq, final SendCallback sendCallback, long timeout)
+SendResult send(final Message msg, final MessageQueueSelector selector, final Object arg)
+void send(final Message msg, final MessageQueueSelector selector, final Object arg, final SendCallback sendCallback)
+void send(final Message msg, final MessageQueueSelector selector, final Object arg, final SendCallback sendCallback, final long timeout)
+void sendOneway(final Message msg, MessageQueueSelector selector, Object arg)
+TransactionSendResult sendMessageInTransaction(Message msg, LocalTransactionExecuter tranExecuter, Object arg)
+SendResult send(final Collection msgs)
+void start()
+void shutdown()
```

图 3-4　MQProducer

2）long searchOffset(final MessageQueue mq, final long timestamp)

根据时间戳从队列中查找其偏移量。

3）long maxOffset(final MessageQueue mq)

查找该消息队列中最大的物理偏移量。

4）long minOffset(final MessageQueue mq)

查找该消息队列中最小物理偏移量。

5）MessageExt viewMessage(final String offsetMsgId)

根据消息偏移量查找消息。

6）QueryResult queryMessage(final String topic, final String key, final int maxNum, final long begin, final long end)

根据条件查询消息。

topic：消息主题。

key：消息索引字段。

maxNum：本次最多取出消息条数。

begin：开始时间。

end：结束时间。

7）MessageExt viewMessage(String topic,String msgId)

根据主题与消息 ID 查找消息。

8）List<MessageQueue> fetchPublishMessageQueues(final String topic)

查找该主题下所有的消息队列。

9）SendResult send(final Message msg)

同步发送消息，具体发送到主题中的哪个消息队列由负载算法决定。

10）SendResult send(final Message msg, final long timeout)

同步发送消息，如果发送超过 timeout 则抛出超时异常。

11）void send(final Message msg, final SendCallback sendCallback)

异步发送消息，sendCallback 参数是消息发送成功后的回调方法。

12）void send(final Message msg, final SendCallback sendCallback, final long timeout)

异步发送消息，如果发送超过 timeout 指定的值，则抛出超时异常。

13）void sendOneway(final Message msg)

单向消息发送，就是不在乎发送结果，消息发送出去后该方法立即返回。

14）SendResult send(final Message msg, final MessageQueue mq)

同步方式发送消息，发送到指定消息队列。

15）void send(final Message msg, final MessageQueue mq, final SendCallback sendCallback)

异步方式发送消息，发送到指定消息队列。

16）void sendOneway(final Message msg, final MessageQueue mq)

单向方式发送消息，发送到指定的消息队列。

17）SendResult send(final Message msg, final MessageQueueSelector selector, final Object arg)

消息发送，指定消息选择算法，覆盖消息生产者默认的消息队列负载。

18）SendResult send(final Collection<Message> msgs, final MessageQueue mq, final long timeout)

同步批量消息发送。

代码清单 3-2　DefaultMQProducer 核心属性

```
private String producerGroup;
private String createTopicKey = MixAll.DEFAULT_TOPIC;
private volatile int defaultTopicQueueNums = 4;
private int sendMsgTimeout = 3000;
private int compressMsgBodyOverHowmuch = 1024 * 4;
private int retryTimesWhenSendFailed = 2;
private int retryTimesWhenSendAsyncFailed = 2;
```

```
private boolean retryAnotherBrokerWhenNotStoreOK = false;
private int maxMessageSize = 1024 * 1024 * 4;  // 4M
```

producerGroup：生产者所属组，消息服务器在回查事务状态时会随机选择该组中任何一个生产者发起事务回查请求。

createTopicKey：默认 topicKey。

defaultTopicQueueNums：默认主题在每一个 Broker 队列数量。

sendMsgTimeout：发送消息默认超时时间，默认 3s。

compressMsgBodyOverHowmuch：消息体超过该值则启用压缩，默认 4K。

retryTimesWhenSendFailed：同步方式发送消息重试次数，默认为 2，总共执行 3 次。

retryTimesWhenSendAsyncFailed：异步方式发送消息重试次数，默认为 2。

retryAnotherBrokerWhenNotStoreOK：消息重试时选择另外一个 Broker 时，是否不等待存储结果就返回，默认为 false。

maxMessageSize：允许发送的最大消息长度，默认为 4M，该值最大值为 2^32-1。

3.3.2 消息生产者启动流程

消息生产者是如何一步一步启动的呢？我们可以从这个类的 DefaultMQProducerImpl 的 start 方法来跟踪，具体细节如下。

代码清单 3-3　DefaultMQProducerImpl#start

```
this.checkConfig();
if (!this.defaultMQProducer.getProducerGroup().equals(
    MixAll.CLIENT_INNER_PRODUCER_GROUP)) {
    this.defaultMQProducer.changeInstanceNameToPID();
}
```

Step1：检查 productGroup 是否符合要求；并改变生产者的 instanceName 为进程 ID。

代码清单 3-4　DefaultMQProducerImpl#start

```
this.mQClientFactory = MQClientManager.getInstance().
    getAndCreateMQClientInstance(this.defaultMQProducer, rpcHook);
```

代码清单 3-5　MQClientManager#getAndCreateMQClientInstance

```
public MQClientInstance getAndCreateMQClientInstance(final ClientConfig
                    clientConfig, RPCHook rpcHook) {
    String clientId = clientConfig.buildMQClientId();
    MQClientInstance instance = this.factoryTable.get(clientId);
    if (null == instance) {
        instance = new MQClientInstance(clientConfig.cloneClientConfig(),
            this.factoryIndexGenerator.getAndIncrement(), clientId, rpcHook);
        MQClientInstance prev = this.factoryTable.putIfAbsent(clientId, instance);
        if (prev != null) {
            instance = prev;
```

```
            log.warn("Returned Previous MQClientInstance for clientId:[{}]",
                clientId);
        } else {
            log.info("Created new MQClientInstance for clientId:[{}]",
                clientId);
        }
    }
    return instance;
}
```

Step2：创建 MQClientInstance 实例。整个 JVM 实例中只存在一个 MQClientManager 实例，维护一个 MQClientInstance 缓存表 ConcurrentMap<String/* clientId */, MQClientInstance> factoryTable =new ConcurrentHashMap<String, MQClientInstance>()，也就是同一个 clientId 只会创建一个 MQClientInstance。代码清单 3-6 是创建 clientId 的方法。

代码清单 3-6　ClientConfig#buildMQClientId

```
public String buildMQClientId() {
    StringBuilder sb = new StringBuilder();
    sb.append(this.getClientIP());
    sb.append("@");
    sb.append(this.getInstanceName());
    if (!UtilAll.isBlank(this.unitName)) {
        sb.append("@");
        sb.append(this.unitName);
    }
    return sb.toString();
}
```

clientId 为客户端 IP+instance+(unitname 可选)，如果在同一台物理服务器部署两个应用程序，应用程序岂不是 clientId 相同，会造成混乱？

为了避免这个问题，如果 instance 为默认值 DEFAULT 的话，RocketMQ 会自动将 instance 设置为进程 ID，这样避免了不同进程的相互影响，但同一个 JVM 中的不同消费者和不同生产者在启动时获取到的 MQClientInstane 实例都是同一个。根据后面的介绍，MQClientInstance 封装了 RocketMQ 网络处理 API，是消息生产者（Producer）、消息消费者（Consumer）与 NameServer、Broker 打交道的网络通道。

代码清单 3-7　DefaultMQProducerImpl#start

```
boolean registerOK = mQClientFactory.registerProducer
    (this.defaultMQProducer.getProducerGroup(), this);
if (!registerOK) {
        this.serviceState = ServiceState.CREATE_JUST;
    throw new MQClientException("The producer group[" +
        this.defaultMQProducer.getProducerGroup()
        + "] has been created before, specify another name please." +
```

```
            FAQUrl.suggestTodo(FAQUrl.GROUP_NAME_DUPLICATE_URL),
                null);
}
```

Step3：向 MQClientInstance 注册，将当前生产者加入到 MQClientInstance 管理中，方便后续调用网络请求、进行心跳检测等。

Step4：启动 MQClientInstance，如果 MQClientInstance 已经启动，则本次启动不会真正执行。MQClientInstance 启动过程将在第 5 章讲解消息消费时有详细的介绍。

3.4 消息发送基本流程

消息发送流程主要的步骤：验证消息、查找路由、消息发送 (包含异常处理机制)。

代码清单 3-8　同步消息发送入口

```
DefaultMQProducer#send
public SendResult send(Message msg) throws MQClientException, RemotingException,
    MQBrokerException, InterruptedException{
    return this.defaultMQProducerImpl.send(msg);
}
DefaultMQProducerImpl#send
public SendResult send(Message msg) throws MQClientException, RemotingException,
        MQBrokerException, InterruptedException {
    return send(msg, this.defaultMQProducer.getSendMsgTimeout());
}
public SendResult send(Message msg,long timeout) throws MQClientException,
        RemotingException, MQBrokerException, InterruptedException {
    return this.sendDefaultImpl(msg, CommunicationMode.SYNC, null, timeout);
}
```

默认消息发送以同步方式发送，默认超时时间为 3s。

本节主要以 SendResult sendMessage(Messsage message) 方法为突破口，窥探一下消息发送的基本实现流程。

3.4.1　消息长度验证

消息发送之前，首先确保生产者处于运行状态，然后验证消息是否符合相应的规范，具体的规范要求是主题名称、消息体不能为空、消息长度不能等于 0 且默认不能超过允许发送消息的最大长度 4M（maxMessageSize=1024 * 1024 * 4）。

3.4.2　查找主题路由信息

消息发送之前，首先需要获取主题的路由信息，只有获取了这些信息我们才知道消息要发送到具体的 Broker 节点。

代码清单 3-9　DefaultMQProducerImpl#tryToFindTopicPublishInfo

```
private TopicPublishInfo tryToFindTopicPublishInfo(final String topic) {
    TopicPublishInfo topicPublishInfo = this.topicPublishInfoTable.get(topic);
    if (null == topicPublishInfo || !topicPublishInfo.ok()) {
        this.topicPublishInfoTable.putIfAbsent(topic, new TopicPublishInfo());
        this.mQClientFactory.updateTopicRouteInfoFromNameServer(topic);
        topicPublishInfo = this.topicPublishInfoTable.get(topic);
    }
    if (topicPublishInfo.isHaveTopicRouterInfo() || topicPublishInfo.ok()) {
        return topicPublishInfo;
    } else {
        this.mQClientFactory.updateTopicRouteInfoFromNameServer(topic, true,
            this.defaultMQProducer);
        topicPublishInfo = this.topicPublishInfoTable.get(topic);
        return topicPublishInfo;
    }
}
```

tryToFindTopicPublishInfo 是查找主题的路由信息的方法。如果生产者中缓存了 topic 的路由信息，如果该路由信息中包含了消息队列，则直接返回该路由信息，如果没有缓存或没有包含消息队列，则向 NameServer 查询该 topic 的路由信息。如果最终未找到路由信息，则抛出异常：无法找到主题相关路由信息异常。先看一下 TopicPublishInfo，如图 3-5 所示。

TopicPublishInfo	TopicRouteData
-private boolean orderTopic = false -private boolean haveTopicRouterInfo = false -private List messageQueueList -private volatile ThreadLocalIndex sendWhichQueue -private TopicRouteData topicRouteData	-private String orderTopicConf -private List queueDatas -private List brokerDatas -private HashMap filterServerTable

图 3-5　RocketMQ TopicPublishInfo

下面我们来一一介绍下 TopicPublishInfo 的属性。

orderTopic：是否是顺序消息。

List<MessageQueue> messageQueueList：该主题队列的消息队列。

sendWhichQueue：每选择一次消息队列，该值会自增 1，如果 Integer.MAX_VALUE，则重置为 0，用于选择消息队列。

List<QueueData> queueData：topic 队列元数据。

List<BrokerData> brokerDatas：topic 分布的 broker 元数据。

HashMap<String/* brokerAdress*/,List<String> /*filterServer*/>：broker 上过滤服务器地址列表。

第一次发送消息时，本地没有缓存 topic 的路由信息，查询 NameServer 尝试获取，如果路由信息未找到，再次尝试用默认主题 DefaultMQProducerImpl#createTopicKey 去查询，如果 BrokerConfig#autoCreateTopicEnable 为 true 时，NameServer 将返回路由信息，如果 autoCreateTopicEnable 为 false 将抛出无法找到 topic 路由异常。代码 MQClientInstance#updateTopicRouteInfoFromNameServer 这个方法的功能是消息生产者更新和维护路由缓存，具体代码如下。

代码清单 3-10　MQClientInstance#updateTopicRouteInfoFromNameServer

```
TopicRouteData topicRouteData;
if (isDefault && defaultMQProducer != null) {
    topicRouteData = this.mQClientAPIImpl.getDefaultTopicRouteInfoFromNameServer
            (defaultMQProducer.getCreateTopicKey(),1000 * 3);
    if (topicRouteData != null) {
        for (QueueData data : topicRouteData.getQueueDatas()) {
            int queueNums = Math.min(defaultMQProducer.getDefaultTopicQueueNums(),
                data.getReadQueueNums());
            data.setReadQueueNums(queueNums);
            data.setWriteQueueNums(queueNums);
        }
    }
} else {
    topicRouteData = this.mQClientAPIImpl.getTopicRouteInfoFromNameServer(topic,
        1000 * 3);
}
```

Step1：如果 isDefault 为 true，则使用默认主题去查询，如果查询到路由信息，则替换路由信息中读写队列个数为消息生产者默认的队列个数 (defaultTopicQueueNums)；如果 isDefault 为 false，则使用参数 topic 去查询；如果未查询到路由信息，则返回 false，表示路由信息未变化。

代码清单 3-11　MQClientInstance#updateTopicRouteInfoFromNameServer

```
TopicRouteData old = this.topicRouteTable.get(topic);
boolean changed = topicRouteDataIsChange(old, topicRouteData);
if (!changed) {
    changed = this.isNeedUpdateTopicRouteInfo(topic);
} else {
    log.info("the topic[{}] route info changed, old[{}] ,new[{}]", topic, old,
        topicRouteData);
}
```

Step2：如果路由信息找到，与本地缓存中的路由信息进行对比，判断路由信息是否发生了改变，如果未发生变化，则直接返回 false。

Step3：更新 MQClientInstance Broker 地址缓存表。

代码清单 3-12　MQClientInstance#updateTopicRouteInfoFromNameServer

```
// Update Pub info
{
```

```
        TopicPublishInfo publishInfo = topicRouteData2TopicPublishInfo(topic,
            topicRouteData);
        publishInfo.setHaveTopicRouterInfo(true);
        Iterator<Entry<String, MQProducerInner>> it = this.producerTable.entrySet()
            .iterator();
        while (it.hasNext()) {
            Entry<String, MQProducerInner> entry = it.next();
            MQProducerInner impl = entry.getValue();
            if (impl != null) {
                impl.updateTopicPublishInfo(topic, publishInfo);
            }
        }
    }
```

Step4：根据 topicRouteData 中的 List<QueueData> 转换成 topicPublishInfo 的 List<MessageQueue> 列表。其具体实现在 topicRouteData2TopicPublishInfo，然后会更新该 MQClientInstance 所管辖的所有消息发送关于 topic 的路由信息。

代码清单 3-13　MQClientInstance#updateTopicRouteInfoFromNameServer

```
        List<QueueData> qds = route.getQueueDatas();
        Collections.sort(qds);
        for (QueueData qd : qds) {
            if (PermName.isWriteable(qd.getPerm())) {
                BrokerData brokerData = null;
                for (BrokerData bd : route.getBrokerDatas()) {
                    if (bd.getBrokerName().equals(qd.getBrokerName())) {
                        brokerData = bd;
                        break;
                    }
                }
                if (null == brokerData) {
                    continue;
                }
                if (!brokerData.getBrokerAddrs().containsKey(MixAll.MASTER_ID)) {
                    continue;
                }
                for (int i = 0; i < qd.getWriteQueueNums(); i++) {
                    MessageQueue mq = new MessageQueue(topic, qd.getBrokerName(), i);
                    info.getMessageQueueList().add(mq);
                }
            }
        }
```

循环遍历路由信息的 QueueData 信息，如果队列没有写权限，则继续遍历下一个 QueueData；根据 brokerName 找到 brokerData 信息，找不到或没有找到 Master 节点，则遍历下一个 QueueData；根据写队列个数，根据 topic+ 序号创建 MessageQueue，填充 topicPublishInfo 的 List<QuueMessage>。完成消息发送的路由查找。

3.4.3 选择消息队列

根据路由信息选择消息队列，返回的消息队列按照 broker、序号排序。举例说明，如果 topicA 在 broker-a,broker-b 上分别创建了 4 个队列，那么返回的消息队列：[{"brokerName":"broker-a","queueId":0},{"brokerName":"broker-a","queueId":1},{"brokerName":"broker-a","queueId":2},{"brokerName":"broker-a","queueId":3},{"brokerName":"broker-b","queueId":0},{"brokerName":"broker-b","queueId":1},{"brokerName":"broker-b","queueId":2},{"brokerName":"broker-b","queueId":3}]，那 RocketMQ 如何选择消息队列呢？

首先消息发送端采用重试机制，由 retryTimesWhenSendFailed 指定同步方式重试次数，异步重试机制在收到消息发送结构后执行回调之前进行重试。由 retryTimesWhenSendAsyncFailed 指定，接下来就是循环执行，选择消息队列、发送消息，发送成功则返回，收到异常则重试。选择消息队列有两种方式。

1）sendLatencyFaultEnable=false，默认不启用 Broker 故障延迟机制。

2）sendLatencyFaultEnable=true，启用 Broker 故障延迟机制。

1. 默认机制

sendLatencyFaultEnable=false，调用 TopicPublishInfo#selectOneMessageQueue。

代码清单 3-14　TopicPublishInfo#selectOneMessageQueue

```java
public MessageQueue selectOneMessageQueue(final String lastBrokerName) {
    if (lastBrokerName == null) {
        return selectOneMessageQueue();
    } else {
        int index = this.sendWhichQueue.getAndIncrement();
        for (int i = 0; i < this.messageQueueList.size(); i++) {
            int pos = Math.abs(index++) % this.messageQueueList.size();
            if (pos < 0)
                pos = 0;
            MessageQueue mq = this.messageQueueList.get(pos);
            if (!mq.getBrokerName().equals(lastBrokerName)) {
                return mq;
            }
        }
        return selectOneMessageQueue();
    }
}
public MessageQueue selectOneMessageQueue() {
    int index = this.sendWhichQueue.getAndIncrement();
    int pos = Math.abs(index) % this.messageQueueList.size();
    if (pos < 0)
        pos = 0;
    return this.messageQueueList.get(pos);
}
```

首先在一次消息发送过程中，可能会多次执行选择消息队列这个方法，lastBrokerName 就是上一次选择的执行发送消息失败的 Broker。第一次执行消息队列选择时，lastBrokerName 为 null，此时直接用 sendWhichQueue 自增再获取值，与当前路由表中消息队列个数取模，返回该位置的 MessageQueue(selectOneMessageQueue() 方法)，如果消息发送再失败的话，下次进行消息队列选择时规避上次 MesageQueue 所在的 Broker，否则还是很有可能再次失败。

该算法在一次消息发送过程中能成功规避故障的 Broker，但如果 Broker 宕机，由于路由算法中的消息队列是按 Broker 排序的，如果上一次根据路由算法选择的是宕机的 Broker 的第一个队列，那么随后的下次选择的是宕机 Broker 的第二个队列，消息发送很有可能会失败，再次引发重试，带来不必要的性能损耗，那么有什么方法在一次消息发送失败后，暂时将该 Broker 排除在消息队列选择范围外呢？或许有朋友会问，Broker 不可用后，路由信息中为什么还会包含该 Broker 的路由信息呢？其实这不难解释：首先，NameServer 检测 Broker 是否可用是有延迟的，最短为一次心跳检测间隔（10s）；其次，NameServer 不会检测到 Broker 宕机后马上推送消息给消息生产者，而是消息生产者每隔 30s 更新一次路由信息，所以消息生产者最快感知 Broker 最新的路由信息也需要 30s。如果能引入一种机制，在 Broker 宕机期间，如果一次消息发送失败后，可以将该 Broker 暂时排除在消息队列的选择范围中。

2. Broker 故障延迟机制

代码清单 3-15　MQFaultStrategy#selectOneMessageQueue

```java
public MessageQueue selectOneMessageQueue(final TopicPublishInfo tpInfo, final
        String lastBrokerName) {
    if (this.sendLatencyFaultEnable) {
        try {
            int index = tpInfo.getSendWhichQueue().getAndIncrement();
            for (int i = 0; i < tpInfo.getMessageQueueList().size(); i++) {
                int pos = Math.abs(index++) % tpInfo.getMessageQueueList().size();
                if (pos < 0)
                    pos = 0;
                MessageQueue mq = tpInfo.getMessageQueueList().get(pos);
                if (latencyFaultTolerance.isAvailable(mq.getBrokerName())) {
                    if (null == lastBrokerName ||
                            mq.getBrokerName().equals(lastBrokerName))
                        return mq;
                }
            }
            final String notBestBroker = latencyFaultTolerance.pickOneAtLeast();
            int writeQueueNums = tpInfo.getQueueIdByBroker(notBestBroker);
            if (writeQueueNums > 0) {
                final MessageQueue mq = tpInfo.selectOneMessageQueue();
                if (notBestBroker != null) {
                    mq.setBrokerName(notBestBroker);
```

```
                        mq.setQueueId(tpInfo.getSendWhichQueue().getAndIncrement() %
                            writeQueueNums);
                    }
                    return mq;
                } else {
                    latencyFaultTolerance.remove(notBestBroker);
                }
            } catch (Exception e) {
                log.error("Error occurred when selecting message queue", e);
            }
            return tpInfo.selectOneMessageQueue();
        }
        return tpInfo.selectOneMessageQueue(lastBrokerName);
    }
```

首先对上述代码进行解读。

1）根据对消息队列进行轮询获取一个消息队列。

2）验证该消息队列是否可用，latencyFaultTolerance.isAvailable(mq.getBrokerName())是关键。

3）如果返回的 MessageQueue 可用，移除 latencyFaultTolerance 关于该 topic 条目，表明该 Broker 故障已经恢复。

Broker 故障延迟机制核心类如图 3-6 所示。

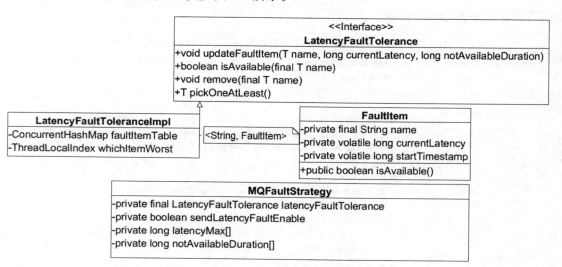

图 3-6　RocketMQ 故障延迟机制核心类

LatencyFaultTolerance：延迟机制接口规范。

1）void updateFaultItem(final T name, final long currentLatency, final long notAvailable-Duration)

更新失败条目。
name：brokerName。
currentLatency：消息发送故障延迟时间。
notAvailableDuration：不可用持续时长，在这个时间内，Broker 将被规避。
2）boolean isAvailable(final T name)
判断 Broker 是否可用。
name：broker 名称。
3）void remove(final T name)
移除 Fault 条目，意味着 Broker 重新参与路由计算。
4）T pickOneAtLeast()
尝试从规避的 Broker 中选择一个可用的 Broker，如果没有找到，将返回 null。
FaultItem：失败条目（规避规则条目）。
1）final String name 条目唯一键，这里为 brokerName。
2）private volatile long currentLatency 本次消息发送延迟。
3）private volatile long startTimestamp 故障规避开始时间。
MQFaultStrategy：消息失败策略，延迟实现的门面类。
1）long[] latencyMax = {50L, 100L, 550L, 1000L, 2000L, 3000L, 15000L}，
2）long[] notAvailableDuration = {0L, 0L, 30000L, 60000L, 120000L, 180000L, 600000L}

latencyMax，根据 currentLatency 本次消息发送延迟，从 latencyMax 尾部向前找到第一个比 currentLatency 小的索引 index，如果没有找到，返回 0。然后根据这个索引从 notAvailableDuration 数组中取出对应的时间，在这个时长内，Broker 将设置为不可用。

下面从源码的角度分析 updateFaultItem、isAvailable 方法的实现原理，如下所示。

代码清单 3-16　DefaultMQProducerImpl#sendDefaultImpl

```
beginTimestampPrev = System.currentTimeMillis();
sendResult = this.sendKernelImpl(msg, mq, communicationMode, sendCallback,
        topicPublishInfo, timeout);
endTimestamp = System.currentTimeMillis();
this.updateFaultItem(mq.getBrokerName(), endTimestamp - beginTimestampPrev,
    false);
```

上述代码如果发送过程中抛出了异常，调用 DefaultMQProducerImpl#updateFaultItem，该方法则直接调用 MQFaultStrategy#updateFaultItem 方法，关注一下各个参数的含义。

第一个参数：broker 名称。

第二个参数：本次消息发送延迟时间 currentLatency。

第三个参数：isolation，是否隔离，该参数的含义如果为 true，则使用默认时长 30s 来计算 Broker 故障规避时长，如果为 false，则使用本次消息发送延迟时间来计算 Broker 故

障规避时长。

代码清单 3-17　MQFaultStrategy#updateFaultItem

```java
public void updateFaultItem(final String brokerName, final long currentLatency,
        boolean isolation) {
    if (this.sendLatencyFaultEnable) {
        long duration = computeNotAvailableDuration(isolation ? 30000 :
            currentLatency);
        this.latencyFaultTolerance.updateFaultItem(brokerName, currentLatency,
            duration);
    }
}
private long computeNotAvailableDuration(final long currentLatency) {
    for (int i = latencyMax.length - 1; i >= 0; i--) {
        if (currentLatency >= latencyMax[i])
            return this.notAvailableDuration[i];
    }
    return 0;
}
```

如果 isolation 为 true，则使用 30s 作为 computeNotAvailableDuration 方法的参数；如果 isolation 为 false，则使用本次消息发送时延作为 computeNotAvailableDuration 方法的参数，那 computeNotAvailableDuration 的作用是计算因本次消息发送故障需要将 Broker 规避的时长，也就是接下来多久的时间内该 Broker 将不参与消息发送队列负载。具体算法：从 latencyMax 数组尾部开始寻找，找到第一个比 currentLatency 小的下标，然后从 notAvailableDuration 数组中获取需要规避的时长，该方法最终调用 LatencyFaultTolerance 的 updateFaultItem。

代码清单 3-18　LatencyFaultToleranceImpl#updateFaultItem

```java
public void updateFaultItem(final String name, final long currentLatency, final
        long notAvailableDuration) {
    FaultItem old = this.faultItemTable.get(name);
    if (null == old) {
        final FaultItem faultItem = new FaultItem(name);
        faultItem.setCurrentLatency(currentLatency);
        faultItem.setStartTimestamp(System.currentTimeMillis() +
            notAvailableDuration);
        old = this.faultItemTable.putIfAbsent(name, faultItem);
        if (old != null) {
            old.setCurrentLatency(currentLatency);
            old.setStartTimestamp(System.currentTimeMillis() +
                notAvailableDuration);
        }
    } else {
        old.setCurrentLatency(currentLatency);
        old.setStartTimestamp(System.currentTimeMillis() +
```

```
            notAvailableDuration);
    }
}
```

根据 broker 名称从缓存表中获取 FaultItem，如果找到则更新 FaultItem，否则创建 FaultItem。这里有两个关键点。

1）currentLatency、startTimeStamp 被 volatile 修饰。

2）startTimeStamp 为当前系统时间加上需要规避的时长。startTimeStamp 是判断 broker 当前是否可用的直接一句，请看 FaultItem#isAvailable 方法。

代码清单 3-19　FaultItem#isAvailable

```
public boolean isAvailable() {
    return (System.currentTimeMillis() - startTimestamp) >= 0;
}
```

3.4.4　消息发送

消息发送 API 核心入口：DefaultMQProducerImpl#sendKernelImpl。

代码清单 3-20　DefaultMQProducerImpl#sendKernelImpl

```
private SendResult sendKernelImpl(final Message msg,
        final MessageQueue mq,
        final CommunicationMode communicationMode,
        final SendCallback sendCallback,
        final TopicPublishInfo topicPublishInfo,
        final long timeout)
```

消息发送参数详解。

1）Message msg：待发送消息。

2）MessageQueue mq：消息将发送到该消息队列上。

3）CommunicationMode communicationMode：消息发送模式，SYNC、ASYNC、ONEWAY。

4）SendCallback sendCallback：异步消息回调函数。

5）TopicPublishInfo topicPublishInfo：主题路由信息。

6）long timeout：消息发送超时时间。

代码清单 3-21　DefaultMQProducerImpl#sendKernelImpl

```
String brokerAddr = this.mQClientFactory.findBrokerAddressInPublish
    (mq.getBrokerName());
if (null == brokerAddr) {
    tryToFindTopicPublishInfo(mq.getTopic());
    brokerAddr = this.mQClientFactory.findBrokerAddressInPublish
        (mq.getBrokerName());
}
```

Step1：根据 MessageQueue 获取 Broker 的网络地址。如果 MQClientInstance 的 brokerAddrTable 未缓存该 Broker 的信息，则从 NameServer 主动更新一下 topic 的路由信息。如果路由更新后还是找不到 Broker 信息，则抛出 MQClientException，提示 Broker 不存在。

代码清单 3-22　DefaultMQProducerImpl#sendKernelImpl

```
//for MessageBatch,ID has been set in the generating process
if (!(msg instanceof MessageBatch)) {
    MessageClientIDSetter.setUniqID(msg);
}
int sysFlag = 0;
if (this.tryToCompressMessage(msg)) {
    sysFlag |= MessageSysFlag.COMPRESSED_FLAG;
}
final String tranMsg = msg.getProperty(MessageConst
        .PROPERTY_TRANSACTION_PREPARED);
if (tranMsg != null && Boolean.parseBoolean(tranMsg)) {
    sysFlag |= MessageSysFlag.TRANSACTION_PREPARED_TYPE;
}
```

Step2：为消息分配全局唯一 ID，如果消息体默认超过 4K(compressMsgBodyOverHowmuch)，会对消息体采用 zip 压缩，并设置消息的系统标记为 MessageSysFlag.COMPRESSED_FLAG。如果是事务 Prepared 消息，则设置消息的系统标记为 MessageSysFlag.TRANSACTION_PREPARED_TYPE。

代码清单 3-23　DefaultMQProducerImpl#sendKernelImpl

```
if (this.hasSendMessageHook()) {
    context = new SendMessageContext();
    context.setProducer(this);
    context.setProducerGroup(this.defaultMQProducer.getProducerGroup());
    context.setCommunicationMode(communicationMode);
    context.setBornHost(this.defaultMQProducer.getClientIP());
    context.setBrokerAddr(brokerAddr);
    context.setMessage(msg);
    context.setMq(mq);
    String isTrans = msg.getProperty(MessageConst.PROPERTY_TRANSACTION_PREPARED);
    if (isTrans != null && isTrans.equals("true")) {
        context.setMsgType(MessageType.Trans_Msg_Half);
    }
    if (msg.getProperty("__STARTDELIVERTIME") != null ||
            msg.getProperty(MessageConst.PROPERTY_DELAY_TIME_LEVEL) != null) {
        context.setMsgType(MessageType.Delay_Msg);
    }
    this.executeSendMessageHookBefore(context);
}
```

Step3：如果注册了消息发送钩子函数，则执行消息发送之前的增强逻辑。通过 Defaul

tMQProducerImpl#registerSendMessageHook 注册钩子处理类，并且可以注册多个。简单看一下钩子处理类接口。

代码清单 3-24　SendMessageHook

```
public interface SendMessageHook {
    String hookName();
    void sendMessageBefore(final SendMessageContext context);
    void sendMessageAfter(final SendMessageContext context);
}
```

代码清单 3-25　DefaultMQProducerImpl#sendKernelImpl

```
SendMessageRequestHeader requestHeader = new SendMessageRequestHeader();
requestHeader.setProducerGroup(this.defaultMQProducer.getProducerGroup());
requestHeader.setTopic(msg.getTopic());
requestHeader.setDefaultTopic(this.defaultMQProducer.getCreateTopicKey());
requestHeader.setDefaultTopicQueueNums(this.defaultMQProducer.getDefaultTopic
        QueueNums());
requestHeader.setQueueId(mq.getQueueId());
requestHeader.setSysFlag(sysFlag);
requestHeader.setBornTimestamp(System.currentTimeMillis());
requestHeader.setFlag(msg.getFlag());
requestHeader.setProperties(MessageDecoder.messageProperties2String(msg.getPr
        operties()));
requestHeader.setReconsumeTimes(0);
requestHeader.setUnitMode(this.isUnitMode());
requestHeader.setBatch(msg instanceof MessageBatch);
if (requestHeader.getTopic().startsWith(MixAll.RETRY_GROUP_TOPIC_PREFIX)) {
    String reconsumeTimes = MessageAccessor.getReconsumeTime(msg);
    if (reconsumeTimes != null) {
        requestHeader.setReconsumeTimes(Integer.valueOf(reconsumeTimes));
        MessageAccessor.clearProperty(msg,
            MessageConst.PROPERTY_RECONSUME_TIME);
    }
    String maxReconsumeTimes = MessageAccessor.getMaxReconsumeTimes(msg);
    if (maxReconsumeTimes != null) {

        requestHeader.setMaxReconsumeTimes(Integer.valueOf(maxReconsumeTimes));
        MessageAccessor.clearProperty(msg,
            MessageConst.PROPERTY_MAX_RECONSUME_TIMES);
    }
}
```

Step4：构建消息发送请求包。主要包含如下重要信息：生产者组、主题名称、默认创建主题 Key、该主题在单个 Broker 默认队列数、队列 ID（队列序号）、消息系统标记（MessageSysFlag）、消息发送时间、消息标记（RocketMQ 对消息中的 flag 不做任何处理，供应用程序使用）、消息扩展属性、消息重试次数、是否是批量消息等。

代码清单 3-26　MQClientAPIImpl#sendMessage

```
public SendResult sendMessage(final String addr,final String brokerName,
    final Message msg,final SendMessageRequestHeader requestHeader,
        final long timeoutMillis,final CommunicationMode communicationMode,
        final SendCallback sendCallback,final TopicPublishInfo topicPublishInfo,
        final MQClientInstance instance,final int retryTimesWhenSendFailed,
        final SendMessageContext context,final DefaultMQProducerImpl producer
) throws RemotingException, MQBrokerException, InterruptedException {
    RemotingCommand request = null;
    if (sendSmartMsg || msg instanceof MessageBatch) {
        SendMessageRequestHeaderV2 requestHeaderV2 =
            SendMessageRequestHeaderV2.createSendMessageRequestHeaderV2(requ
                estHeader);
        request = RemotingCommand.createRequestCommand(msg instanceof
            MessageBatch ? RequestCode.SEND_BATCH_MESSAGE :
            RequestCode.SEND_MESSAGE_V2, requestHeaderV2);
    } else {
        request = RemotingCommand.createRequestCommand
            (RequestCode.SEND_MESSAGE, requestHeader);
    }
    request.setBody(msg.getBody());
    switch (communicationMode) {
        case ONEWAY:
            this.remotingClient.invokeOneway(addr, request, timeoutMillis);
            return null;
        case ASYNC:
            final AtomicInteger times = new AtomicInteger();
            this.sendMessageAsync(addr, brokerName, msg, timeoutMillis,
                request, sendCallback, topicPublishInfo, instance,
                retryTimesWhenSendFailed, times, context, producer);
            return null;
        case SYNC:
            return this.sendMessageSync(addr, brokerName, msg, timeoutMillis,
                request);
        default:
            assert false;
            break;
    }
    return null;
}
```

Step5：根据消息发送方式，同步、异步、单向方式进行网络传输。

代码清单 3-27　DefaultMQProducerImpl#sendKernelImpl

```
if (this.hasSendMessageHook()) {
    context.setSendResult(sendResult);
    this.executeSendMessageHookAfter(context);
}
```

Step6：如果注册了消息发送钩子函数，执行 after 逻辑。注意，就算消息发送过程中发生 RemotingException、MQBrokerException、InterruptedException 时该方法也会执行。

1. 同步发送

MQ 客户端发送消息的入口是 MQClientAPIImpl#sendMessage。请求命令是 RequestCode.SEND_MESSAGE，我们可以找到该命令的处理类：org.apache.rocketmq.broker.processor.SendMessageProcessor。入口方法在 SendMessageProcessor#sendMessage。

代码清单 3-28　AbstractSendMessageProcessor#msgCheck

```
protected RemotingCommand msgCheck(final ChannelHandlerContext ctx,final
        SendMessageRequestHeader requestHeader, final RemotingCommand response) {
    if(!PermName.isWriteable(this.brokerController.getBrokerConfig().getBrokerP
        ermission())&& this.brokerController.getTopicConfigManager().
            isOrderTopic(requestHeader.getTopic())) {
        response.setCode(ResponseCode.NO_PERMISSION);
        response.setRemark("the broker[" +
            this.brokerController.getBrokerConfig().getBrokerIP1()
                + "] sending message is forbidden");
        return response;
    }
    if (!this.brokerController.getTopicConfigManager().
            isTopicCanSendMessage(requestHeader.getTopic())) {
        String errorMsg = "the topic[" + requestHeader.getTopic() + "] is conflict
            with system reserved words.";
        log.warn(errorMsg);
        response.setCode(ResponseCode.SYSTEM_ERROR);
        response.setRemark(errorMsg);
        return response;
    }
    TopicConfig topicConfig =
            this.brokerController.getTopicConfigManager().
            selectTopicConfig(requestHeader.getTopic());
    if (null == topicConfig) {
        int topicSysFlag = 0;
        if (requestHeader.isUnitMode()) {
            if (requestHeader.getTopic().startsWith(
                    MixAll.RETRY_GROUP_TOPIC_PREFIX)) {
                topicSysFlag = TopicSysFlag.buildSysFlag(false, true);
            } else {
                topicSysFlag = TopicSysFlag.buildSysFlag(true, false);
            }
        }
        log.warn("the topic {} not exist, producer: {}", requestHeader.getTopic(),
                ctx.channel().remoteAddress());
        topicConfig = this.brokerController.getTopicConfigManager().
            createTopicInSendMessageMethod(
                requestHeader.getTopic(),requestHeader.getDefaultTopic(),
                RemotingHelper.parseChannelRemoteAddr(ctx.channel()),
```

```java
                            requestHeader.getDefaultTopicQueueNums(), topicSysFlag);
            if (null == topicConfig) {
                if (requestHeader.getTopic().startsWith(
                    MixAll.RETRY_GROUP_TOPIC_PREFIX)) {
                    topicConfig = this.brokerController.getTopicConfigManager().
                        createTopicInSendMessageBackMethod(
                            requestHeader.getTopic(), 1, PermName.PERM_WRITE |
                            PermName.PERM_READ,topicSysFlag);
                }
            }
            if (null == topicConfig) {
                response.setCode(ResponseCode.TOPIC_NOT_EXIST);
                response.setRemark("topic[" + requestHeader.getTopic() + "] not
                    exist, apply first please!"
                    + FAQUrl.suggestTodo(FAQUrl.APPLY_TOPIC_URL));
                return response;
            }
            int queueIdInt = requestHeader.getQueueId();
            int idValid = Math.max(topicConfig.getWriteQueueNums(),
                topicConfig.getReadQueueNums());
            if (queueIdInt >= idValid) {
                String errorInfo = String.format("request queueId[%d] is illegal, %s
                    Producer: %s",
                    queueIdInt,
                    topicConfig.toString(),
                    RemotingHelper.parseChannelRemoteAddr(ctx.channel()));
                log.warn(errorInfo);
                response.setCode(ResponseCode.SYSTEM_ERROR);
                response.setRemark(errorInfo);
                return response;
            }
        return response;
}
```

Step1：检查消息发送是否合理，这里完成了以下几件事情。

1）检查该 Broker 是否有写权限。

2）检查该 Topic 是否可以进行消息发送。主要针对默认主题，默认主题不能发送消息，仅仅供路由查找。

3）在 NameServer 端存储主题的配置信息，默认路径：${ROCKET_HOME}/store/config/topic.json。下面是主题存储信息。order：是否是顺序消息；perm：权限码；readQueueNums：读队列数量；writeQueueNums：写队列数量；topicName：主题名称；topicSysFlag：topic Flag，当前版本暂为保留；topicFilterType：主题过滤方式，当前版本仅支持 SINGLE_TAG。

4）检查队列，如果队列不合法，返回错误码。

Step2：如果消息重试次数超过允许的最大重试次数，消息将进入到 DLD 延迟队列。延迟队列主题：%DLQ%+ 消费组名，延迟队列在消息消费时将重点讲解。

Step3：调用 DefaultMessageStore#putMessage 进行消息存储。关于消息存储的实现细节将在第 4 章重点剖析。

2. 异步发送

消息异步发送是指消息生产者调用发送的 API 后，无须阻塞等待消息服务器返回本次消息发送结果，只需要提供一个回调函数，供消息发送客户端在收到响应结果回调。异步方式相比同步方式，消息发送端的发送性能会显著提高，但为了保护消息服务器的负载压力，RocketMQ 对消息发送的异步消息进行了并发控制，通过参数 clientAsyncSemaphoreValue 来控制，默认为 65535。异步消息发送虽然也可以通过 DefaultMQProducer#retryTimesWhenSendAsyncFailed 属性来控制消息重试次数，但是重试的调用入口是在收到服务端响应包时进行的，如果出现网络异常、网络超时等将不会重试。

3. 单向发送

单向发送是指消息生产者调用消息发送的 API 后，无须等待消息服务器返回本次消息发送结果，并且无须提供回调函数，表示消息发送压根就不关心本次消息发送是否成功，其实现原理与异步消息发送相同，只是消息发送客户端在收到响应结果后什么都不做而已，并且没有重试机制。

3.5 批量消息发送

批量消息发送是将同一主题的多条消息一起打包发送到消息服务端，减少网络调用次数，提高网络传输效率。当然，并不是在同一批次中发送的消息数量越多性能就越好，其判断依据是单条消息的长度，如果单条消息内容比较长，则打包多条消息发送会影响其他线程发送消息的响应时间，并且单批次消息发送总长度不能超过 DefaultMQProducer#maxMessageSize。批量消息发送要解决的是如何将这些消息编码以便服务端能够正确解码出每条消息的消息内容。

那 RocketMQ 如何编码多条消息呢？我们首先梳理一下 RocketMQ 网络请求命令设计。其类图如图 3-7 所示。

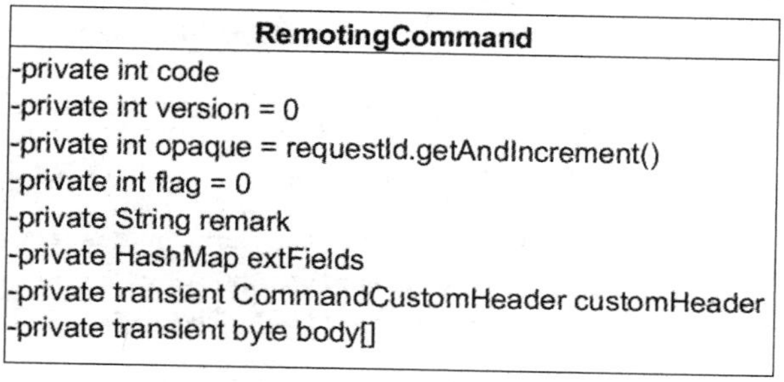

图 3-7 RocketMQ 请求命令类图

下面我们来一一介绍下 RemotingCommand 的属性。

1）code：请求命令编码，请求命令类型。

2）version：版本号。

3）opaque：客户端请求序号。

4）flag：标记。倒数第一位表示请求类型，0：请求；1：返回。倒数第二位，1：表示oneway。

5）remark：描述。

6）extFields：扩展属性。

7）customeHeader：每个请求对应的请求头信息。

8）byte[] body：消息体内容。

单条消息发送时，消息体的内容将保存在 body 中。批量消息发送，需要将多条消息体的内容存储在 body 中，如何存储方便服务端正确解析出每条消息呢？

RocketMQ 采取的方式是，对单条消息内容使用固定格式进行存储，如图 3-8 所示。

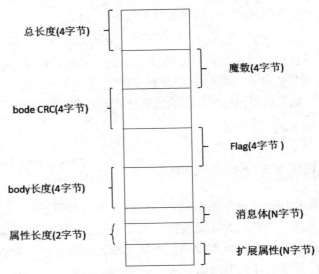

图 3-8　RocetMQ 批量消息封装格式

接下来梳理一下批量消息发送的核心流程。

代码清单 3-29　DefaultMQProducer#send 消息批量发送

```
public SendResult send(Collection<Message> msgs) throws MQClientException,
        RemotingException, MQBrokerException, InterruptedException {
    return this.defaultMQProducerImpl.send(batch(msgs));
}
```

首先在消息发送端，调用 batch 方法，将一批消息封装成 MessageBatch 对象。MessageBatch 继承自 Message 对象，MessageBatch 内部持有 List<Message> messages。这样的话，批量消息发送与单条消息发送的处理流程完全一样。MessageBatch 只需要将该集合中的每条消息的消息体 body 聚合成一个 byte[] 数值,在消息服务端能够从该 byte[] 数值中正确解析出消息即可。

代码清单 3-30　Message'Batch#encode

```
public byte[] encode() {
    return MessageDecoder.encodeMessages(messages);
}
```

在创建 RemotingCommand 对象时将调用 messageBatch#encode() 方法填充到 Remoting-Command 的 body 域中。多条消息编码格式如图 3-8 所示,对应代码见代码清单 3-31。

代码清单 3-31　MessageDecoder#encodeMessage

```
public static byte[] encodeMessage(Message message) {
    //only need flag, body, properties
    byte[] body = message.getBody();
    int bodyLen = body.length;
    String properties = messageProperties2String(message.getProperties());
    byte[] propertiesBytes = properties.getBytes(CHARSET_UTF8);
    //note properties length must not more than Short.MAX
    short propertiesLength = (short) propertiesBytes.length;
    int sysFlag = message.getFlag();
    int storeSize = 4 // 1 TOTALSIZE
        + 4 // 2 MAGICCOD
        + 4 // 3 BODYCRC
        + 4 // 4 FLAG
        + 4 + bodyLen // 4 BODY
        + 2 + propertiesLength;
    ByteBuffer byteBuffer = ByteBuffer.allocate(storeSize);
    // 1 TOTALSIZE
    byteBuffer.putInt(storeSize);
    // 2 MAGICCODE
    byteBuffer.putInt(0);
    // 3 BODYCRC
    byteBuffer.putInt(0);

    // 4 FLAG
    int flag = message.getFlag();
    byteBuffer.putInt(flag);
    // 5 BODY
    byteBuffer.putInt(bodyLen);
    byteBuffer.put(body);
    // 6 properties
    byteBuffer.putShort(propertiesLength);
    byteBuffer.put(propertiesBytes);
    return byteBuffer.array();
}
```

在消息发送端将会按照上述结构进行解码,然后整个发送流程与单个消息发送没什么差异，就不一一介绍了。

3.6 本章小结

本章重点剖析了消息发送的整个流程，重点如下。

1）消息生产者启动流程

重点理解 MQClientInstance、消息生产者之间的关系。

2）消息队列负载机制

消息生产者在发送消息时，如果本地路由表中未缓存 topic 的路由信息，向 NameServer 发送获取路由信息请求，更新本地路由信息表，并且消息生产者每隔 30s 从 NameServer 更新路由表。

3）消息发送异常机制

消息发送高可用主要通过两个手段：重试与 Broker 规避。Broker 规避就是在一次消息发送过程中发现错误，在某一时间段内，消息生产者不会选择该 Broker(消息服务器) 上的消息队列，提高发送消息的成功率。

4）批量消息发送

RocketMQ 支持将同一主题下的多条消息一次性发送到消息服务端。

本章在讨论消息发送流程中并没有深入去跟踪消息是如何存储在消息服务器上的，下一章将重点讲解 RocketMQ 消息存储机制。

第 4 章 RocketMQ 消息存储

目前的 MQ 中间件从存储模型来看，分为需要持久化和不需要持久化的两种模型，现在大多数的 MQ 都是支持持久化存储的，比如 ActiveMQ、RabbitMQ、Kafka,RocketMQ，而 ZeroMQ 却不需要支持持久化存储。然而业务系统也大多需要 MQ 有持久存储的能力，能大大增加系统的高可用性。从存储方式和效率来看，文件系统高于 KV 存储，KV 存储又高于关系型数据库，直接操作文件系统肯定是最快的，但可靠性却是最低的，而关系型数据库的性能和可靠性与文件系统恰恰相反，第 4 章主要分析 RocketMQ 的消息存储机制。

本章重点内容如下。
- RocketMQ 存储概要设计
- 消息发送存储流程
- 存储文件组织与内存映射机制
- RocketMQ 存储文件
- 消息消费队列、索引文件构建机和制
- RocketMQ 文件恢复机制
- RocketMQ 刷盘机制
- RocketMQ 文件删除机制

4.1 存储概要设计

RocketMQ 主要存储的文件包括 Comitlog 文件、ConsumeQueue 文件、IndexFile 文件。RocketMQ 将所有主题的消息存储在同一个文件中，确保消息发送时顺序写文件，尽最大的能力确保消息发送的高性能与高吞吐量。但由于消息中间件一般是基于消息主题的订阅机制，这样便给按照消息主题检索消息带来了极大的不便。为了提高消息消费的效率，RocketMQ 引入了 ConsumeQueue 消息队列文件，每个消息主题包含多个消息消费队列，每一个消息队列有一个消息文件。IndexFile 索引文件，其主要设计理念就是为了加速消息的检索性能，根据消息的属性快速从 Commitlog 文件中检索消息。RocketMQ 是一款高性能的消息中间件，存储部分的设计是核心，存储的核心是 IO 访问性能，本章也会重点剖析 RocketMQ 是如何提高 IO 访问性能的。进入 RocketMQ 存储剖析之前，先看一下 RocketMQ 数据流向，如图 4-1 所示。

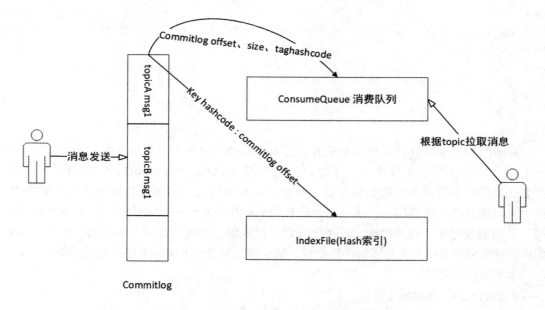

图 4-1　RocketMQ 消息存储设计原理图

1）CommitLog：消息存储文件，所有消息主题的消息都存储在 CommitLog 文件中。
2）ConsumeQueue：消息消费队列，消息到达 CommitLog 文件后，将异步转发到消息消费队列，供消息消费者消费。
3）IndexFile：消息索引文件，主要存储消息 Key 与 Offset 的对应关系。
4）事务状态服务：存储每条消息的事务状态。
5）定时消息服务：每一个延迟级别对应一个消息消费队列，存储延迟队列的消息拉取进度。

4.2　初识消息存储

消息存储实现类：org.apache.rocketmq.store.DefaultMessageStore，它是存储模块里面

最重要的一个类，包含了很多对存储文件操作的 API，其他模块对消息实体的操作都是通过 DefaultMessageStore 进行操作，其类图如图 4-2 所示。

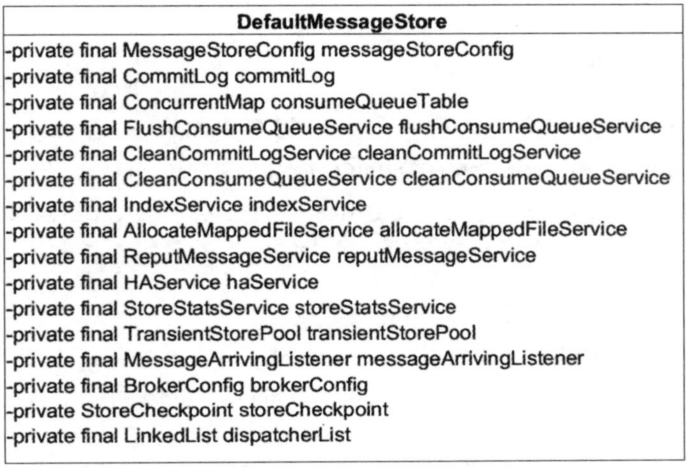

```
DefaultMessageStore
-private final MessageStoreConfig messageStoreConfig
-private final CommitLog commitLog
-private final ConcurrentMap consumeQueueTable
-private final FlushConsumeQueueService flushConsumeQueueService
-private final CleanCommitLogService cleanCommitLogService
-private final CleanConsumeQueueService cleanConsumeQueueService
-private final IndexService indexService
-private final AllocateMappedFileService allocateMappedFileService
-private final ReputMessageService reputMessageService
-private final HAService haService
-private final StoreStatsService storeStatsService
-private final TransientStorePool transientStorePool
-private final MessageArrivingListener messageArrivingListener
-private final BrokerConfig brokerConfig
-private StoreCheckpoint storeCheckpoint
-private final LinkedList dispatcherList
```

图 4-2　DefaultMessageStore 类图

让我们来一一介绍 DefaultMessageStore 的核心属性。

1）MessageStoreConfig messageStoreConfig：消息存储配置属性。

2）CommitLog commitLog：CommitLog 文件的存储实现类。

3）ConcurrentMap<String/* topic */, ConcurrentMap<Integer/* queueId */, ConsumeQueue>> consumeQueueTable：消息队列存储缓存表，按消息主题分组。

4）FlushConsumeQueueService flushConsumeQueueService：消息队列文件 ConsumeQueue 刷盘线程。

5）CleanCommitLogService cleanCommitLogService：清除 CommitLog 文件服务。

6）CleanConsumeQueueService cleanConsumeQueueService：清除 ConsumeQueue 文件服务。

7）IndexService indexService：索引文件实现类。

8）AllocateMappedFileService allocateMappedFileService：MappedFile 分配服务。

9）ReputMessageService reputMessageService：CommitLog 消息分发，根据 CommitLog 文件构建 ConsumeQueue、IndexFile 文件。

10）HAService haService：存储 HA 机制。

11）TransientStorePool transientStorePool：消息堆内存缓存。

12）MessageArrivingListener messageArrivingListener：消息拉取长轮询模式消息达到监听器。

13）BrokerConfig brokerConfig：Broker 配置属性。

14）StoreCheckpoint storeCheckpoint：文件刷盘检测点。

15）LinkedList<CommitLogDispatcher> dispatcherList：CommitLog 文件转发请求。

4.3 消息发送存储流程

本节将以消息发送存储为突破点，一点一点揭开 RocketMQ 存储设计的神秘面纱。消息存储入口：org.apache.rocketmq.store.DefaultMessageStore#putMessage。

Step1：如果当前 Broker 停止工作或 Broker 为 SLAVE 角色或当前 Rocket 不支持写入则拒绝消息写入；如果消息主题长度超过 256 个字符、消息属性长度超过 65536 个字符将拒绝该消息写入。

如果日志中包含 "message store is not writeable, so putMessage is forbidden"，出现这种日志最有可能是磁盘空间不足，在写 ConsumeQueue、IndexFile 文件出现错误时会拒绝消息再次写入。

Step2：如果消息的延迟级别大于 0，将消息的原主题名称与原消息队列 ID 存入消息属性中，用延迟消息主题 SCHEDULE_TOPIC、消息队列 ID 更新原先消息的主题与队列，这是并发消息消费重试关键的一步，下一章会重点探讨消息重试机制与定时消息的实现原理。

代码清单 4-1　CommitLog#putMessage

```
MappedFile unlockMappedFile = null;
MappedFile mappedFile = this.mappedFileQueue.getLastMappedFile();
```

Step3：获取当前可以写入的 Commitlog 文件，RocketMQ 物理文件的组织方式如图 4-3 所示。

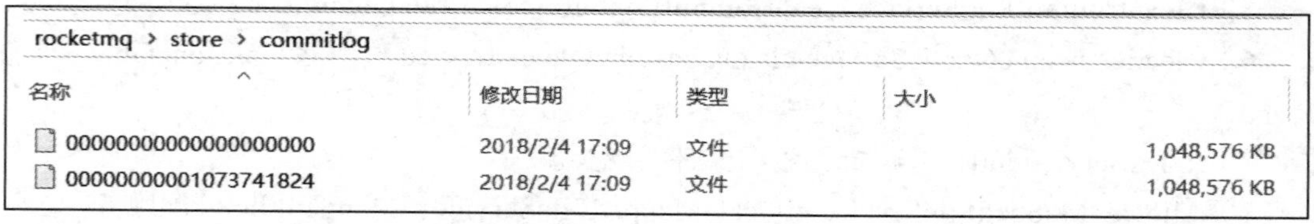

rocketmq > store > commitlog			
名称 ^	修改日期	类型	大小
00000000000000000000	2018/2/4 17:09	文件	1,048,576 KB
00000000001073741824	2018/2/4 17:09	文件	1,048,576 KB

图 4-3　CommitLog 文件组织方式

Commitlog 文件存储目录为 ${ROCKET_HOME}/store/commitlog 目录，每一个文件默认 1G，一个文件写满后再创建另外一个，以该文件中第一个偏移量为文件名，偏移量小于 20 位用 0 补齐。图 4-3 所示的第一个文件初始偏移量为 0，第二个文件的 1073741824，代表该文件中的第一条消息的物理偏移量为 1073741824，这样根据物理偏移量能快速定位到消息。MappedFileQueue 可以看作是 ${ROCKET_HOME}/store/commitlog 文件夹，而 MappedFile 则对应该文件夹下一个个的文件。

Step4：在写入 CommitLog 之前，先申请 putMessageLock，也就是将消息存储到 CommitLog 文件中是串行的。

代码清单 4-2　CommitLog#putMessage

```
messageExtBatch.setStoreTimestamp(beginLockTimestamp);
if (null == mappedFile || mappedFile.isFull()) {
    mappedFile = this.mappedFileQueue.getLastMappedFile(0);
}
if (null == mappedFile) {
    log.error("Create maped file1 error, topic: {} clientAddr: {}",
        messageExtBatch.getTopic(), messageExtBatch.getBornHostString());
    beginTimeInLock = 0;
    return new PutMessageResult(PutMessageStatus.CREATE_MAPEDFILE_FAILED, null);
}
```

Step5：设置消息的存储时间，如果 mappedFile 为空，表明 ${ROCKET_HOME}/store/commitlog 目录下不存在任何文件，说明本次消息是第一次消息发送，用偏移量 0 创建第一个 commit 文件，文件为 00000000000000000000，如果文件创建失败，抛出 CREATE_MAPEDFILE_FAILED，很有可能是磁盘空间不足或权限不够。

代码清单 4-3　MappedFile#appendMessagesInner

```
int currentPos = this.wrotePosition.get();
if (currentPos < this.fileSize) {
    ByteBuffer byteBuffer = writeBuffer != null ? writeBuffer.slice() :
            this.mappedByteBuffer.slice();
    byteBuffer.position(currentPos);
    AppendMessageResult result = null;
    if (messageExt instanceof MessageExtBrokerInner) {
        result = cb.doAppend(this.getFileFromOffset(), byteBuffer, this.fileSize -
                currentPos, (MessageExtBrokerInner) messageExt);
    } else if (messageExt instanceof MessageExtBatch) {
        result = cb.doAppend(this.getFileFromOffset(), byteBuffer, this.fileSize -
                currentPos, (MessageExtBatch) messageExt);
    } else {
        return new AppendMessageResult(AppendMessageStatus.UNKNOWN_ERROR);
    }
    this.wrotePosition.addAndGet(result.getWroteBytes());
    this.storeTimestamp = result.getStoreTimestamp();
    return result;
}
```

Step6：将消息追加到 MappedFile 中。首先先获取 MappedFile 当前写指针，如果 currentPos 大于或等于文件大小则表明文件已写满，抛出 AppendMessageStatus.UNKNOWN_ERROR。如果 currentPos 小于文件大小，通过 slice() 方法创建一个与 MappedFile 的共享内存区，并设置 position 为当前指针。

代码清单 4-4　CommitLog$DefaultAppendMessageCallback#doAppend

```
long wroteOffset = fileFromOffset + byteBuffer.position();
this.resetByteBuffer(hostHolder, 8);
String msgId = MessageDecoder.createMessageId(this.msgIdMemory,
    msgInner.getStoreHostBytes(hostHolder), wroteOffset);
```

Step7：创建全局唯一消息 ID，消息 ID 有 16 字节，消息 ID 组成如图 4-4 所示。

图 4-4　消息 ID 组成

但为了消息 ID 可读性，返回给应用程序的 msgId 为字符类型，可以通过 UtilAll. bytes2string 方法将 msgId 字节数组转换成字符串，通过 UtilAll.string2bytes 方法将 msgId 字符串还原成 16 个字节的字节数组，从而根据提取消息偏移量，可以快速通过 msgId 找到消息内容。

代码清单 4-5　CommitLog$DefaultAppendMessageCallback#doAppend

```
keyBuilder.setLength(0);
keyBuilder.append(msgInner.getTopic());
keyBuilder.append('-');
keyBuilder.append(msgInner.getQueueId());
String key = keyBuilder.toString();
Long queueOffset = CommitLog.this.topicQueueTable.get(key);
if (null == queueOffset) {
    queueOffset = 0L;
    CommitLog.this.topicQueueTable.put(key, queueOffset);
}
```

Step8：获取该消息在消息队列的偏移量。CommitLog 中保存了当前所有消息队列的当前待写入偏移量。

代码清单 4-6　CommitLog#calMsgLength

```
private static int calMsgLength(int bodyLength, int topicLength, int
        propertiesLength) {
    final int msgLen = 4 //TOTALSIZE
        + 4 //MAGICCODE
        + 4 //BODYCRC
        + 4 //QUEUEID
        + 4 //FLAG
        + 8 //QUEUEOFFSET
        + 8 //PHYSICALOFFSET
        + 4 //SYSFLAG
```

```
            + 8 //BORNTIMESTAMP
            + 8 //BORNHOST
            + 8 //STORETIMESTAMP
            + 8 //STOREHOSTADDRESS
            + 4 //RECONSUMETIMES
            + 8 //Prepared Transaction Offset
            + 4 + (bodyLength > 0 ? bodyLength : 0) //BODY
            + 1 + topicLength //TOPIC
            + 2 + (propertiesLength > 0 ? propertiesLength : 0) //propertiesLength
            + 0;
        return msgLen;
}
```

Step9：根据消息体的长度、主题的长度、属性的长度结合消息存储格式计算消息的总长度。

RocketMQ 消息存储格式如下。

1）TOTALSIZE：该消息条目总长度，4 字节。

2）MAGICCODE：魔数，4 字节。固定值 0xdaa320a7。

3）BODYCRC：消息体 crc 校验码，4 字节。

4）QUEUEID：消息消费队列 ID，4 字节。

5）FLAG：消息 FLAG，RocketMQ 不做处理，供应用程序使用，默认 4 字节。

6）QUEUEOFFSET：消息在消息消费队列的偏移量，8 字节。

7）PHYSICALOFFSET：消息在 CommitLog 文件中的偏移量，8 字节。

8）SYSFLAG：消息系统 Flag，例如是否压缩、是否是事务消息等，4 字节。

9）BORNTIMESTAMP：消息生产者调用消息发送 API 的时间戳，8 字节。

10）BORNHOST：消息发送者 IP、端口号，8 字节。

11）STORETIMESTAMP：消息存储时间戳，8 字节。

12）STOREHOSTADDRESS：Broker 服务器 IP+ 端口号，8 字节。

13）RECONSUMETIMES：消息重试次数，4 字节。

14）Prepared Transaction Offset：事务消息物理偏移量，8 字节。

15）BodyLength：消息体长度，4 字节。

16）Body：消息体内容，长度为 bodyLenth 中存储的值。

17）TopicLength：主题存储长度，1 字节，表示主题名称不能超过 255 个字符。

18）Topic：主题，长度为 TopicLength 中存储的值。

19）PropertiesLength：消息属性长度，2 字节，表示消息属性长度不能超过 65536 个字符。

20）Properties：消息属性，长度为 PropertiesLength 中存储的值。

上述表示 CommitLog 条目是不定长的，每一个条目的长度存储在前 4 个字节中。

代码清单 4-7　CommitLog$DefaultAppendMessageCallback#doAppend

```
if ((msgLen + END_FILE_MIN_BLANK_LENGTH) > maxBlank) {
    this.resetByteBuffer(this.msgStoreItemMemory, maxBlank);
    this.msgStoreItemMemory.putInt(maxBlank);
    this.msgStoreItemMemory.putInt(CommitLog.BLANK_MAGIC_CODE);
    final long beginTimeMills = CommitLog.this.defaultMessageStore.now();
    byteBuffer.put(this.msgStoreItemMemory.array(), 0, maxBlank);
    return new AppendMessageResult(AppendMessageStatus.END_OF_FILE, wroteOffset,
        maxBlank, msgId, msgInner.getStoreTimestamp(),queueOffset,
        CommitLog.this.defaultMessageStore.now() - beginTimeMills);
}
```

Step10：如果消息长度 +END_FILE_MIN_BLANK_LENGTH 大于 CommitLog 文件的空闲空间，则返回 AppendMessageStatus.END_OF_FILE，Broker 会重新创建一个新的 CommitLog 文件来存储该消息。从这里可以看出，每个 CommitLog 文件最少会空闲 8 个字节，高 4 字节存储当前文件剩余空间，低 4 字节存储魔数：CommitLog.BLANK_MAGIC_CODE。

代码清单 4-8　CommitLog$DefaultAppendMessageCallback#doAppend

```
final long beginTimeMills = CommitLog.this.defaultMessageStore.now();
// Write messages to the queue buffer
byteBuffer.put(this.msgStoreItemMemory.array(), 0, msgLen);
AppendMessageResult result = new
    AppendMessageResult(AppendMessageStatus.PUT_OK, wroteOffset, msgLen,
        msgId,msgInner.getStoreTimestamp(), queueOffset,
    CommitLog.this.defaultMessageStore.now() - beginTimeMills);
```

Step11：将消息内容存储到 ByteBuffer 中，然后创建 AppendMessageResult。这里只是将消息存储在 MappedFile 对应的内存映射 Buffer 中，并没有刷写到磁盘，追加结果如图 4-5 所示。

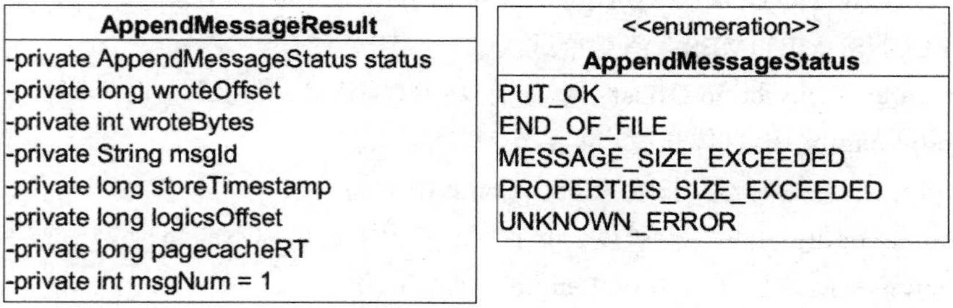

AppendMessageResult	<<enumeration>> AppendMessageStatus
-private AppendMessageStatus status -private long wroteOffset -private int wroteBytes -private String msgId -private long storeTimestamp -private long logicsOffset -private long pagecacheRT -private int msgNum = 1	PUT_OK END_OF_FILE MESSAGE_SIZE_EXCEEDED PROPERTIES_SIZE_EXCEEDED UNKNOWN_ERROR

图 4-5　AppendMessageResult 类图

下面我们来一一介绍下 AppendMessageResult 的属性。

1）AppendMessageStatus status：消息追加结果，取值 PUT_OK：追加成功；END_

OF_FILE：超过文件大小；MESSAGE_SIZE_EXCEEDED：消息长度超过最大允许长度；PROPERTIES_SIZE_EXCEEDED：消息属性超过最大允许长度；UNKNOWN_ERROR：未知异常。

2）long wroteOffset：消息的物理偏移量。

3）String msgId：消息 ID。

4）long storeTimestamp：消息存储时间戳。

5）long logicsOffset：消息消费队列逻辑偏移量，类似于数组下标。

6）long pagecacheRT = 0：当前未使用。

7）int msgNum = 1：消息条数，批量消息发送时消息条数。

代码清单 4-9　CommitLog$DefaultAppendMessageCallback#doAppend

```
case MessageSysFlag.TRANSACTION_NOT_TYPE:
case MessageSysFlag.TRANSACTION_COMMIT_TYPE:
    CommitLog.this.topicQueueTable.put(key, ++queueOffset);
break;
```

Step12：更新消息队列逻辑偏移量。

Step13：处理完消息追加逻辑后将释放 putMessageLock 锁。

代码清单 4-10　CommitLog#putMessage

```
handleDiskFlush(result, putMessageResult, msg);
handleHA(result, putMessageResult, msg);
return putMessageResult;
```

Step14：DefaultAppendMessageCallback#doAppend 只是将消息追加在内存中，需要根据是同步刷盘还是异步刷盘方式，将内存中的数据持久化到磁盘，关于刷盘操作后面会详细介绍。然后执行 HA 主从同步复制，主从同步将在第 7 章详细介绍。

消息发送的基本流程就介绍到这里，下一节开始详细剖析 RocketMQ 消息存储机制的各个方面。

4.4　存储文件组织与内存映射

RocketMQ 通过使用内存映射文件来提高 IO 访问性能，无论是 CommitLog、ConsumeQueue 还是 IndexFile，单个文件都被设计为固定长度，如果一个文件写满以后再创建一个新文件，文件名就为该文件第一条消息对应的全局物理偏移量。例如 CommitLog 文件的组织方式如图 4-6 所示。

rocketmq > store > commitlog			
名称	修改日期	类型	大小
00000000000000000000	2018/2/4 17:09	文件	1,048,576 KB
00000000001073741824	2018/2/4 17:09	文件	1,048,576 KB

图 4-6　CommitLog 文件物理组织方式

RocketMQ 使用 MappedFile、MappedFileQueue 来封装存储文件，其关系如图 4-7 所示。

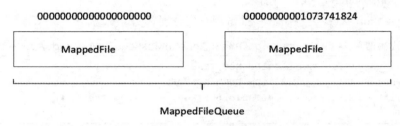

图 4-7　MappedFile、MappedFileQueue 组织方式

4.4.1　MappedFileQueue 映射文件队列

MappedFileQueue 是 MappedFile 的管理容器，MappedFileQueue 是对存储目录的封装，例如 CommitLog 文件的存储路径 ${ROCKET_HOME}/store/commitlog/，该目录下会存在多个内存映射文件 (MappedFile)。MappedFileQueue 类图如图 4-8 所示。

```
MappedFileQueue
-private final String storePath
-private final int mappedFileSize
-private final CopyOnWriteArrayList mappedFiles
-private final AllocateMappedFileService allocateMappedFileService
-private long flushedWhere = 0
-private long committedWhere = 0
-private volatile long storeTimestamp = 0
```

图 4-8　MappedFileQueue 类图

下面让我们一一来介绍 MappedFileQueue 的核心属性。

1）String storePath：存储目录。

2）int mappedFileSize：单个文件的存储大小。

3）CopyOnWriteArrayList<MappedFile> mappedFiles：MappedFile 文件集合。

4）AllocateMappedFileService allocateMappedFileService：创建 MappedFile 服务类。

5）long flushedWhere = 0：当前刷盘指针，表示该指针之前的所有数据全部持久化到磁盘。

6）long committedWhere = 0：当前数据提交指针，内存中 ByteBuffer 当前的写指针，该值大于等于 flushedWhere。

接下来重点分析一下根据不同查询维度查找 MappedFile。

代码清单 4-11　MappedFileQueue#getMappedFileByTime

```
public MappedFile getMappedFileByTime(final long timestamp) {
    Object[] mfs = this.copyMappedFiles(0);
    if (null == mfs)
        return null;
    for (int i = 0; i < mfs.length; i++) {
        MappedFile mappedFile = (MappedFile) mfs[i];
        if (mappedFile.getLastModifiedTimestamp() >= timestamp) {
            return mappedFile;
        }
    }
    return (MappedFile) mfs[mfs.length - 1];}
```

根据消息存储时间戳来查找 MappdFile。从 MappedFile 列表中第一个文件开始查找，找到第一个最后一次更新时间大于待查找时间戳的文件，如果不存在，则返回最后一个 MappedFile 文件。

代码清单 4-12　MappedFileQueue#findMappedFileByOffset

```
public MappedFile findMappedFileByOffset(final long offset, final boolean
        returnFirstOnNotFound) {
    // 省略外层 try ...catch
    MappedFile mappedFile = this.getFirstMappedFile();
    if (mappedFile != null) {
        int index = (int) ((offset / this.mappedFileSize) -
                (mappedFile.getFileFromOffset() / this.mappedFileSize));
        if (index < 0 || index >= this.mappedFiles.size()) {
            // 省略警告日志
        }
        try {
            return this.mappedFiles.get(index);
        } catch (Exception e) {
            if (returnFirstOnNotFound) {
                return mappedFile;
            }
            LOG_ERROR.warn("findMappedFileByOffset failure. ", e);
        }
    }
}
```

根据消息偏移量 offset 查找 MappedFile。根据 offet 查找 MappedFile 直接使用 offset%-mappedFileSize 是否可行？答案是否定的，由于使用了内存映射，只要存在于存储目录下的文件，都需要对应创建内存映射文件，如果不定时将已消费的消息从存储文件中删除，会造成极大的内存压力与资源浪费，所有 RocketMQ 采取定时删除存储文件的策略，也就是说在存储文件中，第一个文件不一定是 00000000000000000000，因为该文件在某一时刻会被删除，故根据 offset 定位 MappedFile 的算法为 (int) ((offset / this.mappedFileSize) - (mappedFile.getFileFromOffset() / this.MappedFileSize))。

代码清单 4-13　MappedFileQueue#getMinOffset

```java
public long getMinOffset() {
    if (!this.mappedFiles.isEmpty()) {
        try {
            return this.mappedFiles.get(0).getFileFromOffset();
        } catch (IndexOutOfBoundsException e) {
        } catch (Exception e) {
            log.error("getMinOffset has exception.", e);
        }
    }
    return -1;
}
```

获取存储文件最小偏移量，从这里也可以看出，并不是直接返回 0，而是返回 MappedFile 的 getFileFormOffset()。

代码清单 4-14　MappedFileQueue#getMaxOffset

```java
public long getMaxOffset() {
    MappedFile mappedFile = getLastMappedFile();
    if (mappedFile != null) {
        return mappedFile.getFileFromOffset() + mappedFile.getReadPosition();
    }
    return 0;
}
```

获取存储文件的最大偏移量。返回最后一个 MappedFile 文件的 fileFromOffset 加上 MappedFile 文件当前的写指针。

代码清单 4-15　MappedFileQueue#getMaxWrotePosition

```java
public long getMaxWrotePosition() {
    MappedFile mappedFile = getLastMappedFile();
    if (mappedFile != null) {
        return mappedFile.getFileFromOffset() + mappedFile.getWrotePosition();
    }
    return 0;
}
```

返回存储文件当前的写指针。返回最后一个文件的 fileFromOffset 加上当前写指针位置。MappedFileQueue 的相关业务方法在具体使用到时再去剖析。

4.4.2 MappedFile 内存映射文件

MappedFile 是 RocketMQ 内存映射文件的具体实现，如图 4-9 所示。

```
                    MappedFile
-public static final int OS_PAGE_SIZE = 1024 * 4
-private static final AtomicLong TOTAL_MAPPED_VIRTUAL_MEMORY
-private static final AtomicInteger TOTAL_MAPPED_FILES
-protected final AtomicInteger wrotePosition
-protected final AtomicInteger committedPosition
-private final AtomicInteger flushedPosition
-protected int fileSize
-protected FileChannel fileChannel
-protected ByteBuffer writeBuffer = null
-protected TransientStorePool transientStorePool
-private String fileName
-private long fileFromOffset
-private File file
-private MappedByteBuffer mappedByteBuffer
-private volatile long storeTimestamp = 0
-private boolean firstCreateInQueue
```

图 4-9　MappedFile 类图

下面让我们一一来介绍 MappedFile 的核心属性。

1）int OS_PAGE_SIZE：操作系统每页大小，默认 4k。

2）AtomicLong TOTAL_MAPPED_VIRTUAL_MEMORY：当前 JVM 实例中 MappedFile 虚拟内存。

3）AtomicInteger TOTAL_MAPPED_FILES：当前 JVM 实例中 MappedFile 对象个数。

4）AtomicInteger wrotePosition：当前该文件的写指针，从 0 开始（内存映射文件中的写指针）。

5）AtomicInteger committedPosition：当前文件的提交指针，如果开启 transientStorePoolEnable，则数据会存储在 TransientStorePool 中，然后提交到内存映射 ByteBuffer 中，再刷写到磁盘。

6）AtomicInteger flushedPosition：刷写到磁盘指针，该指针之前的数据持久化到磁盘中。

7）int fileSize：文件大小。

8）FileChannel fileChannel：文件通道。

9）ByteBuffer writeBuffer：堆内存 ByteBuffer，如果不为空，数据首先将存储在该 Buffer 中，然后提交到 MappedFile 对应的内存映射文件 Buffer。transientStorePoolEnable 为 true 时不为空。

10）TransientStorePool transientStorePool：堆内存池，transientStorePoolEnable 为 true 时启用。

11）String fileName：文件名称。
12）long fileFromOffset：该文件的初始偏移量。
13）File file：物理文件。
14）MappedByteBuffer mappedByteBuffer：物理文件对应的内存映射 Buffer。
15）volatile long storeTimestamp = 0：文件最后一次内容写入时间。
16）boolean firstCreateInQueue：是否是 MappedFileQueue 队列中第一个文件。

1. MappedFile 初始化

根据是否开启 transientStorePoolEnable 存在两种初始化情况。transientStorePoolEnable 为 true 表示内容先存储在堆外内存，然后通过 Commit 线程将数据提交到内存映射 Buffer 中，再通过 Flush 线程将内存映射 Buffer 中的数据持久化到磁盘中。

代码清单 4-16　MappedFile#init(final String fileName, final int fileSize)

```
this.fileName = fileName;
this.fileSize = fileSize;
this.file = new File(fileName);
this.fileFromOffset = Long.parseLong(this.file.getName());
ensureDirOK(this.file.getParent());
this.fileChannel = new RandomAccessFile(this.file, "rw").getChannel();
this.mappedByteBuffer = this.fileChannel.map(MapMode.READ_WRITE, 0, fileSize);
TOTAL_MAPPED_VIRTUAL_MEMORY.addAndGet(fileSize);
TOTAL_MAPPED_FILES.incrementAndGet();
```

初始化 fileFromOffset 为文件名，也就是文件名代表该文件的起始偏移量，通过 RandomAccessFile 创建读写文件通道，并将文件内容使用 NIO 的内存映射 Buffer 将文件映射到内存中。

代码清单 4-17　MappedFile#init

```
public void init(final String fileName, final int fileSize,
        final TransientStorePool transientStorePool) throws IOException {
    init(fileName, fileSize);
    this.writeBuffer = transientStorePool.borrowBuffer();
    this.transientStorePool = transientStorePool;
}
```

如果 transientStorePoolEnable 为 true，则初始化 MappedFile 的 writeBuffer，该 buffer 从 transientStorePool，下一节重点分析一下 TransientStorePool。

2. MappedFile 提交 (commit)

内存映射文件的提交动作由 MappedFile 的 commit 方法实现，如代码清单 4-18 所示。

代码清单 4-18　MappedFile#commit

```
public int commit(final int commitLeastPages) {
    if (writeBuffer == null) {
```

```
            return this.wrotePosition.get();
    }
    if (this.isAbleToCommit(commitLeastPages)) {
        if (this.hold()) {
              commit0(commitLeastPages);
              this.release();
        } else {
              log.warn("in commit, hold failed, commit offset = " +
                  this.committedPosition.get());
        }
    }
    if (writeBuffer != null && this.transientStorePool != null &&
          this.fileSize == this.committedPosition.get()) {
         this.transientStorePool.returnBuffer(writeBuffer);
         this.writeBuffer = null;
    }
    return this.committedPosition.get();
}
```

执行提交操作，commitLeastPages 为本次提交最小的页数，如果待提交数据不满commitLeastPages，则不执行本次提交操作，待下次提交。writeBuffer 如果为空，直接返回wrotePosition 指针，无须执行 commit 操作，表明 commit 操作主体是 writeBuffer。

代码清单 4-19　MappedFile#isAbleToCommit

```
protected boolean isAbleToCommit(final int commitLeastPages) {
    int flush = this.committedPosition.get();
    int write = this.wrotePosition.get();
    if (this.isFull()) {
        return true;
    }
    if (commitLeastPages > 0) {
        return ((write / OS_PAGE_SIZE) - (flush / OS_PAGE_SIZE)) >= commitLeastPages;
    }
    return write > flush;
}
```

判断是否执行 commit 操作。如果文件已满返回 true；如果 commitLeastPages 大于 0，则比较 wrotePosition(当前 writeBuffe 的写指针) 与上一次提交的指针 (committedPosition) 的差值，除以 OS_PAGE_SIZE 得到当前脏页的数量，如果大于 commitLeastPages 则返回 true；如果 commitLeastPages 小于 0 表示只要存在脏页就提交。

代码清单 4-20　MappedFile#commit0

```
protected void commit0(final int commitLeastPages) {
    int writePos = this.wrotePosition.get();
    int lastCommittedPosition = this.committedPosition.get();
    if (writePos - this.committedPosition.get() > 0) {
        try {
```

```
                    ByteBuffer byteBuffer = writeBuffer.slice();
                    byteBuffer.position(lastCommittedPosition);
                    byteBuffer.limit(writePos);
                    this.fileChannel.position(lastCommittedPosition);
                    this.fileChannel.write(byteBuffer);
                    this.committedPosition.set(writePos);
        } catch (Throwable e) {
            log.error("Error occurred when commit data to FileChannel.", e);
        }
    }
}
```

具体的提交实现。首先创建 writeBuffer 的共享缓存区，然后将新创建的 position 回退到上一次提交的位置（committedPosition），设置 limit 为 wrotePosition（当前最大有效数据指针），然后把 commitedPosition 到 wrotePosition 的数据复制（写入）到 FileChannel 中，然后更新 committedPosition 指针为 wrotePosition。commit 的作用就是将 MappedFile#writeBuffer 中的数据提交到文件通道 FileChannel 中。

ByteBuffer 使用技巧：slice() 方法创建一个共享缓存区，与原先的 ByteBuffer 共享内存但维护一套独立的指针 (position、mark、limit)。

3. MappedFile 刷盘 (flush)

刷盘指的是将内存中的数据刷写到磁盘，永久存储在磁盘中，其具体实现由 MappedFile 的 flush 方法实现，如代码清单 4-21 所示。

代码清单 4-21　MappedFile#flush

```
public int flush(final int flushLeastPages) {
    if (this.isAbleToFlush(flushLeastPages)) {
        if (this.hold()) {
            int value = getReadPosition();
            try {
                if (writeBuffer != null || this.fileChannel.position() != 0) {
                    this.fileChannel.force(false);
                } else {
                    this.mappedByteBuffer.force();
                }
            } catch (Throwable e) {
                log.error("Error occurred when force data to disk.", e);
            }
            this.flushedPosition.set(value);
            this.release();
        } else {
            this.flushedPosition.set(getReadPosition());
        }
    }
    return this.getFlushedPosition();
}
```

刷写磁盘，直接调用 mappedByteBuffer 或 fileChannel 的 force 方法将内存中的数据持久化到磁盘，那么 flushedPosition 应该等于 MappedByteBuffer 中的写指针；如果 writeBuffer 不为空，则 flushedPosition 应等于上一次 commit 指针；因为上一次提交的数据就是进入到 MappedByteBuffer 中的数据；如果 writeBuffer 为空，数据是直接进入到 MappedByteBuffer，wrotePosition 代表的是 MappedByteBuffer 中的指针，故设置 flushedPosition 为 wrotePosition。

4. 获取 MappedFile 最大读指针（getReadPosition）

RocketMQ 文件的一个组织方式是内存映射文件，预先申请一块连续的固定大小的内存，需要一套指针标识当前最大有效数据的位置，获取最大有效数据偏移量的方法由 MappedFile 的 getReadPosition 方法实现，如代码清单 4-22 所示。

代码清单 4-22　MappedFile#getReadPosition

```
public int getReadPosition() {
    return this.writeBuffer == null ? this.wrotePosition.get() :
        this.committedPosition.get();
}
```

获取当前文件最大的可读指针。如果 writeBuffer 为空，则直接返回当前的写指针；如果 writeBuffer 不为空，则返回上一次提交的指针。在 MappedFile 设计中，只有提交了的数据（写入到 MappedByteBuffer 或 FileChannel 中的数据）才是安全的数据。

代码清单 4-23　MappedFile#selectMappedBuffer

```
public SelectMappedBufferResult selectMappedBuffer(int pos) {
    int readPosition = getReadPosition();
    if (pos < readPosition && pos >= 0) {
        if (this.hold()) {
            ByteBuffer byteBuffer = this.mappedByteBuffer.slice();
            byteBuffer.position(pos);
            int size = readPosition - pos;
            ByteBuffer byteBufferNew = byteBuffer.slice();
            byteBufferNew.limit(size);
            return new SelectMappedBufferResult(this.fileFromOffset + pos,
                byteBufferNew, size, this);
        }
    }
    return null;
}
```

查找 pos 到当前最大可读之间的数据，由于在整个写入期间都未曾改变 MappedByteBuffer 的指针，所以 mappedByteBuffer.slice() 方法返回的共享缓存区空间为整个 MappedFile，然后通过设置 byteBuffer 的 position 为待查找的值，读取字节为当前可读字节长度，最终返回的 ByteBuffer 的 limit（可读最大长度）为 size。整个共享缓存区的容量为（MappedFile#fileSize-pos），故在操作 SelectMappedBufferResult 不能对包含在里面的 ByteBuffer 调用 flip 方法。

操作 ByteBuffer 时如果使用了 slice() 方法，对其 ByteBuffer 进行读取时一般手动指定 position 与 limit 指针，而不是调用 flip 方法来切换读写状态。

5. MappedFile 销毁 (destory)

MappedFile 文件销毁的实现方法为 public boolean destroy(final long intervalForcibly)，intervalForcibly 表示拒绝被销毁的最大存活时间。

代码清单 4-24　MappedFile#shutdown

```
public void shutdown(final long intervalForcibly) {
    if (this.available) {
        this.available = false;
        this.firstShutdownTimestamp = System.currentTimeMillis();
        this.release();
    } else if (this.getRefCount() > 0) {
        if ((System.currentTimeMillis() - this.firstShutdownTimestamp) >=
                intervalForcibly) {
            this.refCount.set(-1000 - this.getRefCount());
            this.release();
        }
    }
}
```

Step1：关闭 MappedFile。初次调用时 this.available 为 true，设置 available 为 false，并设置初次关闭的时间戳（firstShutdownTimestamp）为当前时间戳，然后调用 release() 方法尝试释放资源，release 只有在引用次数小于 1 的情况下才会释放资源；如果引用次数大于 0，对比当前时间与 firstShutdownTimestamp，如果已经超过了其最大拒绝存活期，每执行一次，将引用数减少 1000，直到引用数小于 0 时通过执行 realse 方法释放资源。

代码清单 4-25　MappedFile#isCleanupOver

```
public boolean isCleanupOver() {
    return this.refCount.get() <= 0 && this.cleanupOver;
}
```

Step2：判断是否清理完成，判断标准是引用次数小于等于 0 并且 cleanupOver 为 true，cleanupOver 为 true 的触发条件是 release 成功将 MappedByteBuffer 资源释放。稍后详细分析 release 方法。

代码清单 4-26　MappedFile#destroy

```
this.fileChannel.close();
log.info("close file channel " + this.fileName + " OK");
```

```
long beginTime = System.currentTimeMillis();
boolean result = this.file.delete();
```

Step3：关闭文件通道，删除物理文件。

在整个 MappedFile 销毁过程，首先需要释放资源，释放资源的前提条件是该 MappedFile 的引用小于等于 0，接下来重点看一下 release 方法的实现原理。

代码清单 4-27　ReferenceResource#release

```
public void release() {
    long value = this.refCount.decrementAndGet();
    if (value > 0)
        return;
    synchronized (this) {
        this.cleanupOver = this.cleanup(value);
    }
}
```

将引用次数减 1，如果引用数小于等于 0，则执行 cleanup 方法，接下来重点分析一下 cleanup 方法的实现。

代码清单 4-28　MappedFile#cleanup

```
public boolean cleanup(final long currentRef) {
    if (this.isAvailable()) {
        return false;
    }
    if (this.isCleanupOver()) {
        return true;
    }
    clean(this.mappedByteBuffer);
    TOTAL_MAPPED_VIRTUAL_MEMORY.addAndGet(this.fileSize * (-1));
    TOTAL_MAPPED_FILES.decrementAndGet();
    log.info("unmap file[REF:" + currentRef + "] " + this.fileName + " OK");
    return true;
}
```

如果 available 为 true，表示 MappedFile 当前可用，无须清理，返回 false；如果资源已经被清除，返回 true；如果是堆外内存，调用堆外内存的 cleanup 方法清除，维护 MappedFile 类变量 TOTAL_MAPPED_VIRTUAL_MEMORY、TOTAL_MAPPED_FILES 并返回 true, 表示 cleanupOver 为 true。

4.4.3 TransientStorePool

TransientStorePool：短暂的存储池。RocketMQ 单独创建一个 MappedByteBuffer 内存缓存池，用来临时存储数据，数据先写入该内存映射中，然后由 commit 线程定时将数据从该内存复制到与目的物理文件对应的内存映射中。RokcetMQ 引入该机制主要的原因是提供一种内存锁定，将当前堆外内存一直锁定在内存中，避免被进程将内存交换到磁盘。TransientStorePool 类图如图 4-10 所示。

```
TransientStorePool
-private final int poolSize
-private final int fileSize
-private final Deque availableBuffers
-private final MessageStoreConfig storeConfig
```

图 4-10　TransientStorePool 类图

下面让我们一一介绍 TransientStorePool 的核心属性。

1）int poolSize：avaliableBuffers 个数，可通过在 broker 中配置文件中设置 transient-StorePoolSize，默认为 5。

2）int fileSize：每个 ByteBuffer 大小，默认为 mapedFileSizeCommitLog，表明 TransientStorePool 为 commitlog 文件服务。

3）Deque<ByteBuffer> availableBuffers：ByteBuffer 容器，双端队列。

代码清单 4-29　TransientStorePool#init

```java
public void init() {
    for (int i = 0; i < poolSize; i++) {
        ByteBuffer byteBuffer = ByteBuffer.allocateDirect(fileSize);
        final long address = ((DirectBuffer) byteBuffer).address();
        Pointer pointer = new Pointer(address);
        LibC.INSTANCE.mlock(pointer, new NativeLong(fileSize));
        availableBuffers.offer(byteBuffer);
    }
}
```

创建 poolSize 个堆外内存，并利用 com.sun.jna.Library 类库将该批内存锁定，避免被置换到交换区，提高存储性能。

4.5　RocketMQ 存储文件

RocketMQ 存储路径为 ${ROCKET_HOME}/store，主要存储文件如图 4-11 所示。

下面让我们一一介绍一下 RocketMQ 主要的存储文件夹。

1）commitlog：消息存储目录。

2）config：运行期间一些配置信息，主要包括下列信息。

　　consumerFilter.json：主题消息过滤信息。

　　consumerOffset.json：集群消费模式消息消费进度。

　　delayOffset.json：延时消息队列拉取进度。

subscriptionGroup.json：消息消费组配置信息。
topics.json：topic 配置属性。

图 4-11　RocketMQ 存储目录

3）consumequeue：消息消费队列存储目录。

4）index：消息索引文件存储目录。

5）abort：如果存在 abort 文件说明 Broker 非正常关闭，该文件默认启动时创建，正常退出之前删除。

6）checkpoint：文件检测点，存储 commitlog 文件最后一次刷盘时间戳、consumequeue 最后一次刷盘时间、index 索引文件最后一次刷盘时间戳。

4.5.1　Commitlog 文件

commitlog 目录的组织方式在 4.4 节中已经详细介绍过了，该目录下的文件主要存储消息，其特点是每一条消息长度不相同，消息存储协议已在 4.3 节中详细描述，Commitlog 文件存储的逻辑视图如图 4-12 所示，每条消息的前面 4 个字节存储该条消息的总长度。

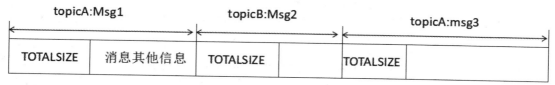

图 4-12　消息组织方式

Commitlog 文件的存储目录默认为 ${ROCKET_HOME}/store/commitlog，可以通过在 broker 配置文件中设置 storePathRootDir 属性来改变默认路径。commitlog 文件默认大小为 1G，可通过在 broker 配置文件中设置 mapedFileSizeCommitLog 属性来改变默认大小。

本节将基于上述存储结构重点分析消息的查找实现，其他诸如文件刷盘、文件恢复机制等将在下文中详细介绍。

代码清单 4-30　Commitlog#getMinOffset

```java
public long getMinOffset() {
    MappedFile mappedFile = this.mappedFileQueue.getFirstMappedFile();
    if (mappedFile != null) {
        if (mappedFile.isAvailable()) {
            return mappedFile.getFileFromOffset();
        } else {
            return this.rollNextFile(mappedFile.getFileFromOffset());
        }
    }
    return -1;
}
```

获取当前 Commitlog 目录最小偏移量，首先获取目录下的第一个文件，如果该文件可用，则返回该文件的起始偏移量，否则返回下一个文件的起始偏移量。

代码清单 4-31　Commitlog#rollNextFile

```java
public long rollNextFile(final long offset) {
    int mappedFileSize = this.defaultMessageStore.getMessageStoreConfig()
            .getMapedFileSizeCommitLog();
    return offset + mappedFileSize - offset % mappedFileSize;
}
```

根据该 offset 返回下一个文件的起始偏移量。首先获取一个文件的大小，减去 (offset % mappedFileSize) 其目的是回到下一文件的起始偏移量。

代码清单 4-32　Commitlog#getMessage

```java
public SelectMappedBufferResult getMessage(final long offset, final int size) {
    int mappedFileSize =
    this.defaultMessageStore.getMessageStoreConfig().
            getMapedFileSizeCommitLog();
    MappedFile mappedFile = this.mappedFileQueue.findMappedFileByOffset(offset,
            offset == 0);
    if (mappedFile != null) {
        int pos = (int) (offset % mappedFileSize);
        return mappedFile.selectMappedBuffer(pos, size);
    }
    return null;
}
```

根据偏移量与消息长度查找消息。首先根据偏移找到所在的物理偏移量，然后用 offset 与文件长度取余得到在文件内的偏移量，从该偏移量读取 size 长度的内容返回即可。如果只根据消息偏移查找消息，则首先找到文件内的偏移量，然后尝试读取 4 个字节获取消息的实际长度，最后读取指定字节即可。

4.5.2 ConsumeQueue 文件

RocketMQ 基于主题订阅模式实现消息消费，消费者关心的是一个主题下的所有消息，但由于同一主题的消息不连续地存储在 commitlog 文件中，试想一下如果消息消费者直接从消息存储文件 (commitlog) 中去遍历查找订阅主题下的消息，效率将极其低下，RocketMQ 为了适应消息消费的检索需求，设计了消息消费队列文件 (Consumequeue)，该文件可以看成是 Commitlog 关于消息消费的"索引"文件，consumequeue 的第一级目录为消息主题，第二级目录为主题的消息队列，如图 4-13 所示。

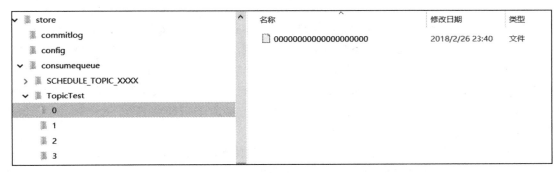

图 4-13　消息消费队列文件组织

为了加速 ConsumeQueue 消息条目的检索速度与节省磁盘空间，每一个 Consumequeue 条目不会存储消息的全量信息，其存储格式如图 4-14 所示。

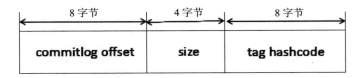

图 4-14　ConsumeQueue 条目

单个 ConsumeQueue 文件中默认包含 30 万个条目，单个文件的长度为 30w × 20 字节，单个 ConsumeQueue 文件可以看出是一个 ConsumeQueue 条目的数组，其下标为 ConsumeQueue 的逻辑偏移量，消息消费进度存储的偏移量即逻辑偏移量。ConsumeQueue 即为 Commitlog 文件的索引文件，其构建机制是当消息到达 Commitlog 文件后，由专门的线程产生消息转发任务，从而构建消息消费队列文件与下文提到的索引文件。

本节只分析如何根据消息逻辑偏移量、时间戳查找消息的实现，下一节将重点讨论消息消费队列的构建、恢复等。

代码清单 4-33　ConsumeQueue#getIndexBuffer

```
public SelectMappedBufferResult getIndexBuffer(final long startIndex) {
    int mappedFileSize = this.mappedFileSize;
```

```
            long offset = startIndex * CQ_STORE_UNIT_SIZE;
            if (offset >= this.getMinLogicOffset()) {
                MappedFile mappedFile = this.mappedFileQueue.findMappedFileByOffset(offset);
                if (mappedFile != null) {
                    SelectMappedBufferResult result = mappedFile.selectMappedBuffer((int)
                        (offset % mappedFileSize));
                    return result;
                }
            }
            return null;
        }
```

根据 startIndex 获取消息消费队列条目。首先 startIndex*20 得到在 consumequeue 中的物理偏移量，如果该 offset 小于 minLogicOffset，则返回 null，说明该消息已被删除；如果大于 minLogicOffset，则根据偏移量定位到具体的物理文件，然后通过 offset 与物理文大小取模获取在该文件的偏移量，从而从偏移量开始连续读取 20 个字节即可。

ConsumeQueue 提供了根据消息存储时间来查找具体实现的算法 getOffsetInQueue-ByTime(final long timestamp)，其具体实现如下。

Step1：首先根据时间戳定位到物理文件，其具体实现在前面有详细介绍，就是从第一个文件开始找到第一个文件更新时间大于该时间戳的文件。

代码清单 4-34　ConsumeQueue#getOffsetInQueueByTime

```
int low = minLogicOffset > mappedFile.getFileFromOffset() ? (int) (minLogicOffset
    - mappedFile.getFileFromOffset()) : 0;
int high = 0;
int midOffset = -1, targetOffset = -1, leftOffset = -1, rightOffset = -1;
long leftIndexValue = -1L, rightIndexValue = -1L;
long minPhysicOffset = this.defaultMessageStore.getMinPhyOffset();
SelectMappedBufferResult sbr = mappedFile.selectMappedBuffer(0);
    if (null != sbr) {
        ByteBuffer byteBuffer = sbr.getByteBuffer();
        high = byteBuffer.limit() - CQ_STORE_UNIT_SIZE;
```

Step2：采用二分查找来加速检索。首先计算最低查找偏移量，取消息队列最小偏移量与该文件最小偏移量二者中的最小偏移量为 low。获取当前存储文件中有效的最小消息物理偏移量 minPhysicOffset，如果查找到消息偏移量小于该物理偏移量，则结束该查找过程。

代码清单 4-35　ConsumeQueue#getOffsetInQueueByTime

```
while (high >= low) {
    midOffset = (low + high) / (2 * CQ_STORE_UNIT_SIZE) * CQ_STORE_UNIT_SIZE;
    byteBuffer.position(midOffset);
    long phyOffset = byteBuffer.getLong();
    int size = byteBuffer.getInt();
    if (phyOffset < minPhysicOffset) {
        low = midOffset + CQ_STORE_UNIT_SIZE;
        leftOffset = midOffset;
```

```
                continue;
            }
            long storeTime =
                    this.defaultMessageStore.getCommitLog().
                        pickupStoreTimestamp(phyOffset, size);
            if (storeTime < 0) {
                return 0;
            } else if (storeTime == timestamp) {
                targetOffset = midOffset;
                break;
            } else if (storeTime > timestamp) {
                high = midOffset - CQ_STORE_UNIT_SIZE;
                rightOffset = midOffset;
                rightIndexValue = storeTime;
            } else {
                low = midOffset + CQ_STORE_UNIT_SIZE;
                leftOffset = midOffset;
                leftIndexValue = storeTime;
            }
        }
```

二分查找的常规退出循环为（low>high），首先查找中间的偏移量 midOffset，将整个 Consume Queue 文件对应的 ByteBuffer 定位到 midOffset，然后读取 4 个字节获取该消息的物理偏移量 offset。

1）如果得到的物理偏移量小于当前的最小物理偏移量，说明待查找的物理偏移量肯定大于 midOffset，所以将 low 设置为 midOffset，然后继续折半查找。

2）如果 offset 大于最小物理偏移量，说明该消息是有效消息，则根据消息偏移量和消息长度获取消息的存储时间戳。

3）如果存储时间小于 0，消息为无效消息，直接返回 0。

4）如果存储时间戳等于待查找时间戳，说明查找到匹配消息，设置 targetOffset 并跳出循环。

5）如果存储时间戳大于待查找时间戳，说明待查找信息小于 midOffset，则设置 high 为 midOffset，并设置 rightIndexValue 等于 midOffset。

6）如果存储时间小于待查找时间戳，说明待查找消息在大于 midOffset，则设置 low 为 midOffset，并设置 leftIndexValue 等于 midOffset。

代码清单 4-36　ConsumeQueue#getOffsetInQueueByTime

```
        if (targetOffset != -1) {
            offset = targetOffset;
        } else {
            if (leftIndexValue == -1) {
                offset = rightOffset;
            } else if (rightIndexValue == -1) {
                offset = leftOffset;
```

```
        } else {
            offset = Math.abs(timestamp - leftIndexValue) > Math.abs(timestamp
                - rightIndexValue) ? rightOffset : leftOffset;
        }
    }
    return (mappedFile.getFileFromOffset() + offset) / CQ_STORE_UNIT_SIZE;
```

Step3：如果 targetOffset 不等于 -1 表示找到了存储时间戳等于待查找时间戳的消息；如果 leftIndexValue 等于 -1，表示返回当前时间戳大并且最接近待查找的偏移量；如果 rightIndexValue 等于 -1，表示返回的消息比待查找时间戳小并且最接近查找的偏移量。

代码清单 4-37　ConsumeQueue#rollNextFile

```
public long rollNextFile(final long index) {
    int mappedFileSize = this.mappedFileSize;
    int totalUnitsInFile = mappedFileSize / CQ_STORE_UNIT_SIZE;
    return index + totalUnitsInFile - index % totalUnitsInFile;
}
```

根据当前偏移量获取下一个文件的起始偏移量。首先获取一个文件包含多少个消息消费队列条目，减去 index%totalUnitsInFile 的目的是选中下一个文件的起始偏移量。

4.5.3　Index 索引文件

消息消费队列是 RocketMQ 专门为消息订阅构建的索引文件，提高根据主题与消息队列检索消息的速度，另外 RocketMQ 引入了 Hash 索引机制为消息建立索引，HashMap 的设计包含两个基本点：Hash 槽与 Hash 冲突的链表结构。RocketMQ 索引文件布局如图 4-15 所示。

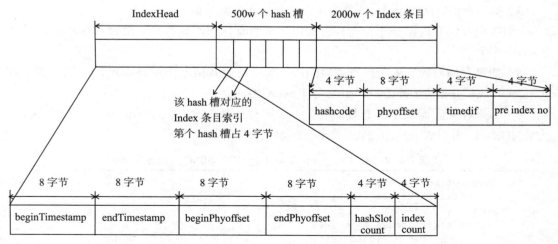

图 4-15　Index 索引文件

从图中可以看出，IndexFile 总共包含 IndexHeader、Hash 槽、Hash 条目（数据）。

1）IndexHeader 头部，包含 40 个字节，记录该 IndexFile 的统计信息，其结构如下。

beginTimestamp：该索引文件中包含消息的最小存储时间。

endTimestamp：该索引文件中包含消息的最大存储时间。

beginPhyoffset：该索引文件中包含消息的最小物理偏移量（commitlog 文件偏移量）。

endPhyoffset：该索引文件中包含消息的最大物理偏移量（commitlog 文件偏移量）。

hashslotCount：hashslot 个数，并不是 hash 槽使用的个数，在这里意义不大。

indexCount：Index 条目列表当前已使用的个数，Index 条目在 Index 条目列表中按顺序存储。

2）Hash 槽，一个 IndexFile 默认包含 500 万个 Hash 槽，每个 Hash 槽存储的是落在该 Hash 槽的 hashcode 最新的 Index 的索引。

3）Index 条目列表，默认一个索引文件包含 2000 万个条目，每一个 Index 条目结构如下。

hashcode：key 的 hashcode。

phyoffset：消息对应的物理偏移量。

timedif：该消息存储时间与第一条消息的时间戳的差值，小于 0 该消息无效。

preIndexNo：该条目的前一条记录的 Index 索引，当出现 hash 冲突时，构建的链表结构。

接下来将重点分析如何将 Map<String/* 消息索引 key*/,long phyOffset/* 消息物理偏移量 */> 存入索引文件，以及如何根据消息索引 key 快速查找消息。

RocketMQ 将消息索引键与消息偏移量映射关系写入到 IndexFile 的实现方法为：public boolean putKey（final String key, final long phyOffset, final long storeTimestamp），参数含义分别为消息索引、消息物理偏移量、消息存储时间。

代码清单 4-38　IndexFile#putKey

```
if (this.indexHeader.getIndexCount() < this.indexNum) {
    int keyHash = indexKeyHashMethod(key);
    int slotPos = keyHash % this.hashSlotNum;
    int absSlotPos = IndexHeader.INDEX_HEADER_SIZE + slotPos * hashSlotSize;
    // ...
}
```

Step1：如果当前已使用条目大于等于允许最大条目数时，则返回 fasle，表示当前索引文件已写满。如果当前索引文件未写满则根据 key 算出 key 的 hashcode，然后 keyHash 对 hash 槽数量取余定位到 hashcode 对应的 hash 槽下标，hashcode 对应的 hash 槽的物理地址为 IndexHeader 头部 (40 字节) 加上下标乘以每个 hash 槽的大小 (4 字节)。

> 注意　根据定位 hash 槽算法，如果不同 key 的 hashcode 值相同或不同的 key 不同的 hashcode 但对 hash 槽数量取余后结果相同都将引发 Hash 冲突，那 IndxFile 如何解决这个问题呢？

代码清单 4-39　IndexFile#putKey

```
int slotValue = this.mappedByteBuffer.getInt(absSlotPos);
if (slotValue <= invalidIndex || slotValue > this.indexHeader.getIndexCount())
{
    slotValue = invalidIndex;
}
```

Step2：读取 hash 槽中存储的数据，如果 hash 槽存储的数据小于 0 或大于当前索引文件中的索引条目格式，则将 slotValue 设置为 0。

代码清单 4-40　IndexFile#putKey

```
long timeDiff = storeTimestamp - this.indexHeader.getBeginTimestamp();
timeDiff = timeDiff / 1000;
if (this.indexHeader.getBeginTimestamp() <= 0) {
    timeDiff = 0;
} else if (timeDiff > Integer.MAX_VALUE) {
    timeDiff = Integer.MAX_VALUE;
} else if (timeDiff < 0) {
    timeDiff = 0;
}
```

Step3：计算待存储消息的时间戳与第一条消息时间戳的差值，并转换成秒。

代码清单 4-41　IndexFile#putKey

```
int absIndexPos =  IndexHeader.INDEX_HEADER_SIZE + this.hashSlotNum *
        hashSlotSize + this.indexHeader.getIndexCount() * indexSize;
this.mappedByteBuffer.putInt(absIndexPos, keyHash);
this.mappedByteBuffer.putLong(absIndexPos + 4, phyOffset);
this.mappedByteBuffer.putInt(absIndexPos + 4 + 8, (int) timeDiff);
this.mappedByteBuffer.putInt(absIndexPos + 4 + 8 + 4, slotValue);
this.mappedByteBuffer.putInt(absSlotPos, this.indexHeader.getIndexCount());
```

Step4：将条目信息存储在 IndexFile 中。

1）计算新添加条目的起始物理偏移量，等于头部字节长度 + hash 槽数量 * 单个 hash 槽大小 (4 个字节) + 当前 Index 条目个数 * 单个 Index 条目大小 (20 个字节)。

2）依次将 hashcode、消息物理偏移量、消息存储时间戳与索引文件时间戳、当前 Hash 槽的值存入 MappedByteBuffer 中。

3）将当前 Index 中包含的条目数量存入 Hash 槽中，将覆盖原先 Hash 槽的值。

这里是 Hash 冲突链式解决方案的关键实现，Hash 槽中存储的是该 HashCode 所对应的最新的 Index 条目的下标，新的 Index 条目的最后 4 个字节存储该 HashCode 上一个条目的 Index 下标。如果 Hash 槽中存储的值为 0 或大于当前 IndexFile 最大条目数或小于 −1，表示该 Hash 槽当前并没有与之对应的 Index 条目。值得关注的是，IndexFile 条目中存储的不是消息索引 key 而是消息属性 key 的 HashCode，在根据 key 查找时需要根据消息物理偏移量找到消息进而再验证消息 key 的值，之所以只存储 HashCode 而不存储具体的 key，是为了将 Index 条目设计为定长结构，才能方便地检索与定位条目。

代码清单4-42　IndexFile#putKey

```
if (this.indexHeader.getIndexCount() <= 1) {
    this.indexHeader.setBeginPhyOffset(phyOffset);
    this.indexHeader.setBeginTimestamp(storeTimestamp);
}
this.indexHeader.incHashSlotCount();
this.indexHeader.incIndexCount();
this.indexHeader.setEndPhyOffset(phyOffset);
this.indexHeader.setEndTimestamp(storeTimestamp);
```

Step5：更新文件索引头信息。如果当前文件只包含一个条目，更新beginPhyOffset与beginTimestamp、更新endPyhOffset、endTimestamp、当前文件使用索引条目等信息。

RocketMQ根据索引key查找消息的实现方法为：selectPhyOffset(List<Long> phyOffsets, String key, int maxNum,long begin, long end)，其参数说如下。

List<Long> phyOffsets：查找到的消息物理偏移量。

String key：索引key。

int maxNum：本次查找最大消息条数。

long begin：开始时间戳。

long end：结束时间戳。

代码清单4-43　IndexFile#selectPhyOffset

```
int keyHash = indexKeyHashMethod(key);
int slotPos = keyHash % this.hashSlotNum;
int absSlotPos = IndexHeader.INDEX_HEADER_SIZE + slotPos * hashSlotSize;
```

Step1：根据key算出key的hashcode，然后keyHash对hash槽数量取余定位到hashcode对应的hash槽下标，hashcode对应的hash槽的物理地址为IndexHeader头部（40字节）加上下标乘以每个hash槽的大小（4字节）。

代码清单4-44　IndexFile#selectPhyOffset

```
int slotValue = this.mappedByteBuffer.getInt(absSlotPos);
if (slotValue <= invalidIndex || slotValue > this.indexHeader.getIndexCount()
        || this.indexHeader.getIndexCount() <= 1) { // return;
}
```

Step2：如果对应的Hash槽中存储的数据小于1或大于当前索引条目个数则表示该HashCode没有对应的条目，直接返回。

代码清单4-45　IndexFile#selectPhyOffset

```
for (int nextIndexToRead = slotValue; ; ){
// ...
}
```

Step3：由于会存在hash冲突，根据slotValue定位该hash槽最新的一个Item条目，将存储的物理偏移加入到 phyOffsets 中,然后继续验证 Item 条目中存储的上一个 Index 下标,如果大于等于 1 并且小于最大条目数,则继续查找,否则结束查找。

代码清单 4-46　IndexFile#selectPhyOffset

```
int absIndexPos = IndexHeader.INDEX_HEADER_SIZE + this.hashSlotNum *
        hashSlotSize + nextIndexToRead * indexSize;
int keyHashRead = this.mappedByteBuffer.getInt(absIndexPos);
long phyOffsetRead = this.mappedByteBuffer.getLong(absIndexPos + 4);
long timeDiff = (long) this.mappedByteBuffer.getInt(absIndexPos + 4 + 8);
int prevIndexRead = this.mappedByteBuffer.getInt(absIndexPos + 4 + 8 + 4);
```

Step4:根据 Index 下标定位到条目的起始物理偏移量,然后依次读取 hashcode、物理偏移量、时间差、上一个条目的 Index 下标。

```
if (timeDiff < 0) {
    break;
}
timeDiff *= 1000L;
long timeRead = this.indexHeader.getBeginTimestamp() + timeDiff;
boolean timeMatched = (timeRead >= begin) && (timeRead <= end);
if (keyHash == keyHashRead && timeMatched) {
        phyOffsets.add(phyOffsetRead);
}
if (prevIndexRead <= invalidIndex
        || prevIndexRead > this.indexHeader.getIndexCount()
        || prevIndexRead == nextIndexToRead || timeRead < begin) {
    break;
}
nextIndexToRead = prevIndexRead;
```

Step5:如果存储的时间差小于 0,则直接结束;如果 hashcode 匹配并且消息存储时间介于待查找时间 start、end 之间则将消息物理偏移量加入到 phyOffsets,并验证条目的前一个 Index 索引,如果索引大于等于 1 并且小于 Index 条目数,则继续查找,否则结束整个查找。

4.5.4　checkpoint 文件

checkpoint 的作用是记录 Comitlog、ConsumeQueue、Index 文件的刷盘时间点,文件固定长度为 4k,其中只用该文件的前面 24 个字节,其存储格式如图 4-16 所示。

图 4-16　checkpoint 文件

physicMsgTimestamp:commitlog 文件刷盘时间点。

logicsMsgTimestamp：消息消费队列文件刷盘时间点。
indexMsgTimestamp：索引文件刷盘时间点。

4.6 实时更新消息消费队列与索引文件

消息消费队列文件、消息属性索引文件都是基于 CommitLog 文件构建的，当消息生产者提交的消息存储在 Commitlog 文件中，ConsumeQueue、IndexFile 需要及时更新，否则消息无法及时被消费，根据消息属性查找消息也会出现较大延迟。RocketMQ 通过开启一个线程 ReputMessageServcie 来准实时转发 CommitLog 文件更新事件，相应的任务处理器根据转发的消息及时更新 ConsumeQueue、IndexFile 文件。

代码清单 4-47　DefaultMessageStore#start

```
if (this.getMessageStoreConfig().isDuplicationEnable()) {
        this.reputMessageService.setReputFromOffset(this.commitLog.
getConfirmOffset());
} else {
        this.reputMessageService.setReputFromOffset(this.commitLog.getMaxOffset());
}
this.reputMessageService.start();
```

Broker 服务器在启动时会启动 ReputMessageService 线程，并初始化一个非常关键的参数 reputFfomOffset，该参数的含义是 ReputMessageService 从哪个物理偏移量开始转发消息给 ConsumeQueue 和 IndexFile。如果允许重复转发，reputFromOffset 设置为 CommitLog 的提交指针；如果不允许重复转发，reputFromOffset 设置为 Commitlog 的内存中最大偏移量。

代码清单 4-48　DefaultMessageStore#run

```
public void run() {
    DefaultMessageStore.log.info(this.getServiceName() + " service started");
    while (!this.isStopped()) {
        try {
            Thread.sleep(1);
            this.doReput();
        } catch (Exception e) {
            DefaultMessageStore.log.warn(this.getServiceName() + " service has
                exception. ", e);
        }
    }
    DefaultMessageStore.log.info(this.getServiceName() + " service end");
}
```

ReputMessageService 线程每执行一次任务推送休息 1 毫秒就继续尝试推送消息到消息消费队列和索引文件，消息消费转发的核心实现在 doReput 方法中实现。

代码清单 4-49　DefaultMessageStore#doReput

```
SelectMappedBufferResult result = 
        DefaultMessageStore.this.commitLog.getData(reputFromOffset);
```

Step1：返回 reputFromOffset 偏移量开始的全部有效数据 (commitlog 文件)。然后循环读取每一条消息。

代码清单 4-50　DefaultMessageStore#doReput

```
DispatchRequest dispatchRequest = DefaultMessageStore.this.commitLog.
        checkMessageAndReturnSize(result.getByteBuffer(), false, false);
int size = dispatchRequest.getMsgSize();
if (dispatchRequest.isSuccess()) {
    if (size > 0) {
            DefaultMessageStore.this.doDispatch(dispatchRequest);
    }
}
```

Step2：从 result 返回的 ByteBuffer 中循环读取消息，一次读取一条，创建 DispatchRequest 对象。DispatchRequest 类图如图 4-17 所示，如果消息长度大于 0，则调用 doDispatch 方法。最终将分别调用 CommitLogDispatcherBuildConsumeQueue（构建消息消费队列）、CommitLogDispatcherBuildIndex（构建索引文件）。

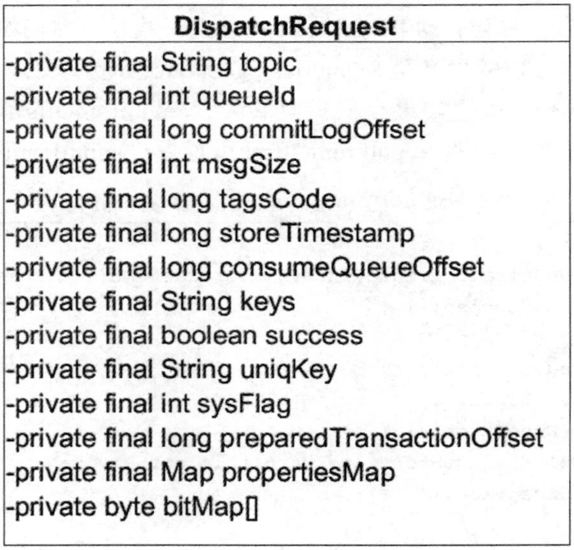

```
DispatchRequest
-private final String topic
-private final int queueId
-private final long commitLogOffset
-private final int msgSize
-private final long tagsCode
-private final long storeTimestamp
-private final long consumeQueueOffset
-private final String keys
-private final boolean success
-private final String uniqKey
-private final int sysFlag
-private final long preparedTransactionOffset
-private final Map propertiesMap
-private byte bitMap[]
```

图 4-17　DispatchRequest 类图

下面让我们一一介绍 DispatchRequest 的核心属性。
1）String topic：消息主题名称。
2）int queueId：消息队列 ID。

3）long commitLogOffset：消息物理偏移量。
4）int msgSize：消息长度。
5）long tagsCode：消息过滤 tag hashcode。
6）long storeTimestamp：消息存储时间戳。
7）long consumeQueueOffset：消息队列偏移量。
8）String keys：消息索引 key。多个索引 key 用空格隔开，例如"key1 key2"。
9）boolean success：是否成功解析到完整的消息。
10）String uniqKey：消息唯一键。
11）int sysFlag：消息系统标记。
12）long preparedTransactionOffset：消息预处理事务偏移量。
13）Map<String, String> propertiesMap：消息属性。
14）byte[] bitMap：位图。

4.6.1 根据消息更新 ConumeQueue

消息消费队列转发任务实现类为：CommitLogDispatcherBuildConsumeQueue，内部最终将调用 putMessagePositionInfo 方法。

代码清单 4-51　DefaultMessageStore#putMessagePositionInfo

```
public void putMessagePositionInfo(DispatchRequest dispatchRequest) {
    ConsumeQueue cq = this.findConsumeQueue(dispatchRequest.getTopic(),
        dispatchRequest.getQueueId());
    cq.putMessagePositionInfoWrapper(dispatchRequest);
}
```

Step1：根据消息主题与队列 ID，先获取对应的 ConumeQueue 文件，其逻辑比较简单，因为每一个消息主题对应一个消息消费队列目录，然后主题下每一个消息队列对应一个文件夹，然后取出该文件夹最后的 ConsumeQueue 文件即可。

代码清单 4-52　ConsumeQueue#putMessagePositionInfo

```
this.byteBufferIndex.flip();
this.byteBufferIndex.limit(CQ_STORE_UNIT_SIZE);
this.byteBufferIndex.putLong(offset);
this.byteBufferIndex.putInt(size);
this.byteBufferIndex.putLong(tagsCode);
final long expectLogicOffset = cqOffset * CQ_STORE_UNIT_SIZE;
MappedFile mappedFile = this.mappedFileQueue.getLastMappedFile
        (expectLogicOffset);
if (mappedFile != null) {
    return mappedFile.appendMessage(this.byteBufferIndex.array());
}
```

Step2：依次将消息偏移量、消息长度、tag hashcode 写入到 ByteBuffer 中，并根据

consumeQueueOffset 计算 ConumeQueue 中的物理地址，将内容追加到 ConsumeQueue 的内存映射文件中（本操作只追击并不刷盘），ConumeQueue 的刷盘方式固定为异步刷盘模式。

4.6.2 根据消息更新 Index 索引文件

Hash 索引文件转发任务实现类：CommitLogDispatcherBuildIndex。

代码清单 4-53　CommitLogDispatcherBuildIndex#dispatch

```
public void dispatch(DispatchRequest request) {
    if (DefaultMessageStore.this.messageStoreConfig.isMessageIndexEnable()) {
        DefaultMessageStore.this.indexService.buildIndex(request);
    }
}
```

如果 messsageIndexEnable 设置为 true，则调用 IndexService#buildIndex 构建 Hash 索引，否则忽略本次转发任务。

代码清单 4-54　IndexService#buildIndex

```
IndexFile indexFile = retryGetAndCreateIndexFile();
if (indexFile != null) {
    long endPhyOffset = indexFile.getEndPhyOffset();
    DispatchRequest msg = req;
    String topic = msg.getTopic();
    String keys = msg.getKeys();
    if (msg.getCommitLogOffset() < endPhyOffset) {
        return;
    }
    // 省略部分代码...
}
```

Step1：获取或创建 IndexFile 文件并获取所有文件最大的物理偏移量。如果该消息的物理偏移量小于索引文件中的物理偏移，则说明是重复数据，忽略本次索引构建。

代码清单 4-55　IndexService#buildIndex

```
if (req.getUniqKey() != null) {
    indexFile = putKey(indexFile, msg, buildKey(topic, req.getUniqKey()));
    if (indexFile == null) {
        log.error("putKey error commitlog {} uniqkey {}", req.getCommitLogOffset(),
            req.getUniqKey());
        return;
    }
}
```

Step2：如果消息的唯一键不为空，则添加到 Hash 索引中，以便加速根据唯一键检索消息。

代码清单4-56　IndexService#buildIndex

```
if (keys != null && keys.length() > 0) {
    String[] keyset = keys.split(MessageConst.KEY_SEPARATOR);
    for (int i = 0; i < keyset.length; i++) {
        String key = keyset[i];
        if (key.length() > 0) {
            indexFile = putKey(indexFile, msg, buildKey(topic, key));
            // return topic + "#" + key
            if (indexFile == null) {
                log.error("putKey error commitlog {} uniqkey {}",
                    req.getCommitLogOffset(), req.getUniqKey());
                return;
            }
        }
    }
}
```

Step3：构建索引键，RocketMQ 支持为同一个消息建立多个索引，多个索引键空格分开。具体如何构建 Hash 索引在 4.5 节中已做了详细分析。

4.7　消息队列与索引文件恢复

由于 RocketMQ 存储首先将消息全量存储在 Commitlog 文件中，然后异步生成转发任务更新 ConsumeQueue、Index 文件。如果消息成功存储到 Commitlog 文件中，转发任务未成功执行，此时消息服务器 Broker 由于某个原因宕机，导致 Commitlog、ConsumeQueue、IndexFile 文件数据不一致。如果不加以人工修复的话，会有一部分消息即便在 Commitlog 文件中存在，但由于并没有转发到 Consumequeue，这部分消息将永远不会被消费者消费。那 RocketMQ 是如何使 Commitlog、消息消费队列（ConsumeQueue）达到最终一致性的呢？下面详细分析一下 RocketMQ 关于存储文件的加载流程来一窥端倪。

代码清单4-57　DefaultMessageStore#load

```
boolean lastExitOK = !this.isTempFileExist();
private boolean isTempFileExist() {
    String fileName = StorePathConfigHelper.getAbortFile
        (this.messageStoreConfig.getStorePathRootDir());
    File file = new File(fileName);
    return file.exists();
}
```

Step1：判断上一次退出是否正常。其实现机制是 Broker 在启动时创建 ${ROCKET_HOME}/store/abort 文件，在退出时通过注册 JVM 钩子函数删除 abort 文件。如果下一次启动时存在 abort 文件。说明 Broker 是异常退出的，Commitlog 与 Consumequeue 数据有可能不一致，需要进行修复。

代码清单 4-58　DefaultMessageStore#load

```
if (null != scheduleMessageService) {
    result = result && this.scheduleMessageService.load();
}
```

Step2：加载延迟队列，RocketMQ 定时消息相关，该部分将在第 5 章详细分析。

代码清单 4-59　MappedFileQueue#load

```
Arrays.sort(files);
for (File file : files) {
    if (file.length() != this.mappedFileSize) {
            return true;
    }
    try {
      MappedFile mappedFile = new MappedFile(file.getPath(), mappedFileSize);
         mappedFile.setWrotePosition(this.mappedFileSize);
      mappedFile.setFlushedPosition(this.mappedFileSize);
      mappedFile.setCommittedPosition(this.mappedFileSize);
      this.mappedFiles.add(mappedFile);
      log.info("load " + file.getPath() + " OK");
    } catch (IOException e) {
      log.error("load file " + file + " error", e);
      return false;
    }
}
```

Step3：加载 Commitlog 文件，加载 ${ROCKET_HOME}/store/commitlog 目录下所有文件并按照文件名排序。如果文件大小与配置文件的单个文件大小不一致，将忽略该目录下所有文件，然后创建 MappedFile 对象。注意 load 方法将 wrotePosition、flushedPosition、committedPosition 三个指针都设置为文件大小。

Step4：加载消息消费队列，调用 DefaultMessageStore#loadConsumeQueue，其思路与 CommitLog 大体一致，遍历消息消费队列根目录，获取该 Broker 存储的所有主题，然后遍历每个主题目录，获取该主题下的所有消息消费队列，然后分别加载每个消息消费队列下的文件，构建 ConsumeQueue 对象，主要初始化 ConsumeQueue 的 topic、queueId、storePath、mappedFileSize 属性。

代码清单 4-60　DefaultMessageStore#load

```
this.storeCheckpoint = new StoreCheckpoint(StorePathConfigHelper.
    getStoreCheckpoint(this.messageStoreConfig.getStorePathRootDir()));
```

Step5：加载存储检测点，检测点主要记录 commitlog 文件、Consumequeue 文件、Index 索引文件的刷盘点，将在下文的文件刷盘机制中再次提交。

代码清单 4-61　IndexService#load

```
for (File file : files) {
    IndexFile f = new IndexFile(file.getPath(), this.hashSlotNum,
```

```
                    this.indexNum, 0, 0);
    f.load();
       if (!lastExitOK) {
          if (f.getEndTimestamp() > this.defaultMessageStore.getStoreCheckpoint()
                         .getIndexMsgTimestamp()) {
             f.destroy(0);
             continue;
          }
       }
    }
}// 省略了异常代码
```

Step6：加载索引文件，如果上次异常退出，而且索引文件上次刷盘时间小于该索引文件最大的消息时间戳该文件将立即销毁。

代码清单4-62　DefaultMessageStore#recover

```
private void recover(final boolean lastExitOK) {
    this.recoverConsumeQueue();
    if (lastExitOK) {
        this.commitLog.recoverNormally();
    } else {
        this.commitLog.recoverAbnormally();
    }
    this.recoverTopicQueueTable();
}
```

Step7：根据Broker是否是正常停止执行不同的恢复策略，下文将分别介绍异常停止、正常停止的文件恢复机制。

代码清单4-63　DefaultMessageStore#recoverTopicQueueTable

```
private void recoverTopicQueueTable() {
    HashMap<String/* topic-queueid */, Long/* offset */> table = new
            HashMap<String, Long>(1024);
    long minPhyOffset = this.commitLog.getMinOffset();
    for (ConcurrentMap<Integer, ConsumeQueue> maps :
                this.consumeQueueTable.values()) {
        for (ConsumeQueue logic : maps.values()) {
            String key = logic.getTopic() + "-" + logic.getQueueId();
            table.put(key, logic.getMaxOffsetInQueue());
            logic.correctMinOffset(minPhyOffset);
        }
    }
    this.commitLog.setTopicQueueTable(table);
}
```

Step8：恢复ConsumeQueue文件后，将在CommitLog实例中保存每个消息消费队列当前的存储逻辑偏移量，这也是消息中不仅存储主题、消息队列ID还存储了消息队列偏移量的关键所在。

4.7.1 Broker 正常停止文件恢复

Broker 正常停止文件恢复的实现为 CommitLog#recoverNormally。

代码清单 4-64　CommitLog#recoverNormally

```
boolean checkCRCOnRecover =
        this.defaultMessageStore.getMessageStoreConfig().isCheckCRCOnRecover();
final List<MappedFile> mappedFiles = this.mappedFileQueue.getMappedFiles();
if (!mappedFiles.isEmpty()) {
    int index = mappedFiles.size() - 3;
    if (index < 0)
        index = 0;
    // ... 省略部分代码
}
```

Step1：Broker 正常停止再重启时，从倒数第三个文件开始进行恢复，如果不足 3 个文件，则从第一个文件开始恢复。checkCRCOnRecover 参数设置在进行文件恢复时查找消息时是否验证 CRC。

代码清单 4-65　CommitLog#recoverNormally

```
MappedFile mappedFile = mappedFiles.get(index);
ByteBuffer byteBuffer = mappedFile.sliceByteBuffer();
long processOffset = mappedFile.getFileFromOffset();
long mappedFileOffset = 0;
```

Step2：解释一下两个局部变量，mappedFileOffset 为当前文件已校验通过的 offset，processOffset 为 Commitlog 文件已确认的物理偏移量等于 mappedFile.getFileFromOffset 加上 mappedFileOffset。

代码清单 4-66　CommitLog#recoverNormally

```
DispatchRequest dispatchRequest = this.checkMessageAndReturnSize(byteBuffer,
        checkCRCOnRecover);
int size = dispatchRequest.getMsgSize();
if (dispatchRequest.isSuccess() && size > 0) {
    mappedFileOffset += size;
} else if (dispatchRequest.isSuccess() && size == 0) {
    index++;
    if (index >= mappedFiles.size()) {
        break;
    } else {
        mappedFile = mappedFiles.get(index);
        byteBuffer = mappedFile.sliceByteBuffer();
        processOffset = mappedFile.getFileFromOffset();
        mappedFileOffset = 0;
        log.info("recover next physics file, " + mappedFile.getFileName());
    }
} else if (!dispatchRequest.isSuccess()) {
    log.info("recover physics file end, " + mappedFile.getFileName());
```

```
        break;
    }
```

Step3：遍历 Commitlog 文件，每次取出一条消息，如果查找结果为 true 并且消息的长度大于 0 表示消息正确，mappedFileOffset 指针向前移动本条消息的长度；如果查找结果为 true 并且消息的长度等于 0，表示已到该文件的末尾，如果还有下一个文件，则重置 processOffset、mappedFileOffset 重复步骤 3，否则跳出循环；如果查找结构为 false，表明该文件未填满所有消息，跳出循环，结束遍历文件。

代码清单 4-67　CommitLog#recoverNormally

```
processOffset += mappedFileOffset;
this.mappedFileQueue.setFlushedWhere(processOffset);
this.mappedFileQueue.setCommittedWhere(processOffset);
```

Step4：更新 MappedFileQueue 的 flushedWhere 与 commiteedWhere 指针。

代码清单 4-68　MappedFileQueue#truncateDirtyFiles

```
public void truncateDirtyFiles(long offset) {
    List<MappedFile> willRemoveFiles = new ArrayList<MappedFile>();
    for (MappedFile file : this.mappedFiles) {
        long fileTailOffset = file.getFileFromOffset() + this.mappedFileSize;
        if (fileTailOffset > offset) {
            if (offset >= file.getFileFromOffset()) {
                file.setWrotePosition((int) (offset % this.mappedFileSize));
                file.setCommittedPosition((int) (offset % this.mappedFileSize));
                file.setFlushedPosition((int) (offset % this.mappedFileSize));
            } else {
                file.destroy(1000);
                willRemoveFiles.add(file);
            }
        }
    }
    this.deleteExpiredFile(willRemoveFiles);
}
```

Step5：删除 offset 之后的所有文件。遍历目录下的文件，如果文件的尾部偏移量小于 offset 则跳过该文件，如果尾部的偏移量大于 offset，则进一步比较 offset 与文件的开始偏移量，如果 offset 大于文件的起始偏移量，说明当前文件包含了有效偏移里，设置 MappedFile 的 flushedPosition 和 commitedPosition；如果 offset 小于文件的起始偏移量，说明该文件是有效文件后面创建的，调用 MappedFile#destory 释放 MappedFile 占用的内存资源（内存映射与内存通道等），然后加入到待删除文件列表中，最终调用 deleteExpiredFile 将文件从物理磁盘删除。过期文件的删除将在下文详细介绍。

4.7.2 Broker 异常停止文件恢复

Broker 异常停止文件恢复的实现为 CommitLog#recoverAbnormally。异常文件恢复的步骤与正常停止文件恢复的流程基本相同，其主要差别有两个。首先，正常停止默认从倒数第三个文件开始进行恢复，而异常停止则需要从最后一个文件往前走，找到第一个消息存储正常的文件。其次，如果 commitlog 目录没有消息文件，如果在消息消费队列目录下存在文件，则需要销毁。

如何判断一个消息文件是一个正确的文件呢？

代码清单 4-69　CommitLog#isMappedFileMatchedRecover

```
int magicCode = byteBuffer.getInt(MessageDecoder.MESSAGE_MAGIC_CODE_POSTION);
if (magicCode != MESSAGE_MAGIC_CODE) {
    return false;
}
```

Step1：首先判断文件的魔数，如果不是 MESSAGE_MAGIC_CODE，返回 false，表示该文件不符合 commitlog 消息文件的存储格式。

代码清单 4-70　CommitLog#isMappedFileMatchedRecover

```
long storeTimestamp = byteBuffer.getLong(
        MessageDecoder.MESSAGE_STORE_TIMESTAMP_POSTION);
if (0 == storeTimestamp) {
    return false;
}
```

Step2：如果文件中第一条消息的存储时间等于 0，返回 false，说明该消息存储文件中未存储任何消息。

代码清单 4-71　CommitLog#isMappedFileMatchedRecover

```
if (this.defaultMessageStore.getMessageStoreConfig().isMessageIndexEnable()
    && this.defaultMessageStore.getMessageStoreConfig().isMessageIndexSafe()) {
    if (storeTimestamp <=
            this.defaultMessageStore.getStoreCheckpoint().getMinTimestampIndex()) {
        return true;
    }
} else if (storeTimestamp <=
            this.defaultMessageStore.getStoreCheckpoint().getMinTimestamp()) {
        return true;
}
```

Step3：对比文件第一条消息的时间戳与检测点，文件第一条消息的时间戳小于文件检测点说明该文件部分消息是可靠的，则从该文件开始恢复。文件检测点中保存了 Commitlog 文件、消息消费队列（ConsumeQueue）、索引文件（IndexFile）的文件刷盘点，RocketMQ 默认选择这消息文件与消息消费队列这两个文件的时间刷盘点中最小值与消息文件第一消息的时间戳对比，如果 messageIndexEnable 为 true，表示索引文件的刷盘时间点也参与计算。

Step4：如果根据前 3 步算法找到 MappedFile，则遍历 MappedFile 中的消息，验证消息的合法性，并将消息重新转发到消息消费队列与索引文件，该步骤在 4.7.1 节中已详细说明。

Step5：如果未找到有效 MappedFile，则设置 commitlog 目录的 flushedWhere、committedWhere 指针都为 0，并销毁消息消费队列文件。

代码清单 4-72　ConsumeQueue#destroy

```
public void destroy() {
    this.maxPhysicOffset = -1;
    this.minLogicOffset = 0;
    this.mappedFileQueue.destroy();
    if (isExtReadEnable()) {
        this.consumeQueueExt.destroy();
    }
}
```

重置 ConsumeQueue 的 maxPhysicOffset 与 minLogicOffset，然后调用 MappedFileQueue 的 destory 方法将消息消费队列目录下的所有文件全部删除。

存储启动时所谓的文件恢复主要完成 flushedPosition、committedWhere 指针的设置、消息消费队列最大偏移量加载到内存，并删除 flushedPosition 之后所有的文件。如果 Broker 异常启动，在文件恢复过程中，RocketMQ 会将最后一个有效文件中的所有消息重新转发到消息消费队列与索引文件，确保不丢失消息，但同时会带来消息重复的问题，纵观 RocktMQ 的整体设计思想，RocketMQ 保证消息不丢失但不保证消息不会重复消费，故消息消费业务方需要实现消息消费的幂等设计。

4.8　文件刷盘机制

RocketMQ 的存储与读写是基于 JDK NIO 的内存映射机制（MappedByteBuffer）的，消息存储时首先将消息追加到内存，再根据配置的刷盘策略在不同时间进行刷写磁盘。如果是同步刷盘，消息追加到内存后，将同步调用 MappedByteBuffer 的 force（）方法；如果是异步刷盘，在消息追加到内存后立刻返回给消息发送端。RocketMQ 使用一个单独的线程按照某一个设定的频率执行刷盘操作。通过在 broker 配置文件中配置 flushDiskType 来设定刷盘方式，可选值为 ASYNC_FLUSH（异步刷盘）、SYNC_FLUSH（同步刷盘），默认为异步刷盘。本书默认以消息存储文件 Commitlog 文件刷盘机制为例来剖析 RocketMQ 的刷盘机制，ConsumeQueue、IndexFile 刷盘的实现原理与 Commitlog 刷盘机制类似。RocketMQ 处理刷盘的实现方法为 Commitlog#handleDiskFlush() 方法，刷盘流程作为消息发送、消息存储的子流程，请先重点了解 4.3 节中关于消息存储流程的相关知识。值得注意的是索引文件的刷盘并不是采取定时刷盘机制，而是每更新一次索引文件就会将上一次的改动刷写到磁盘。

4.8.1 Broker 同步刷盘

同步刷盘，指的是在消息追加到内存映射文件的内存中后，立即将数据从内存刷写到磁盘文件，由 CommitLog 的 handleDiskFlush 方法实现，如代码清单 4-73 所示。

代码清单 4-73　CommitLog#handleDiskFlush

```
final GroupCommitService service = (GroupCommitService)
    this.flushCommitLogService;
GroupCommitRequest request = new GroupCommitRequest(result.getWroteOffset() +
        result.getWroteBytes());
service.putRequest(request);
boolean flushOK = request.waitForFlush(this.defaultMessageStore.
        getMessageStoreConfig().getSyncFlushTimeout());
if(!flushOK){
    putMessageResult.setPutMessageStatus(PutMessageStatus.FLUSH_DISK_TIMEOUT);
}
```

同步刷盘实现流程如下。

1）构建 GroupCommitRequest 同步任务并提交到 GroupCommitRequest。

2）等待同步刷盘任务完成，如果超时则返回刷盘错误，刷盘成功后正常返回给调用方。GroupCommitRequest 的类图如图 4-18 所示。

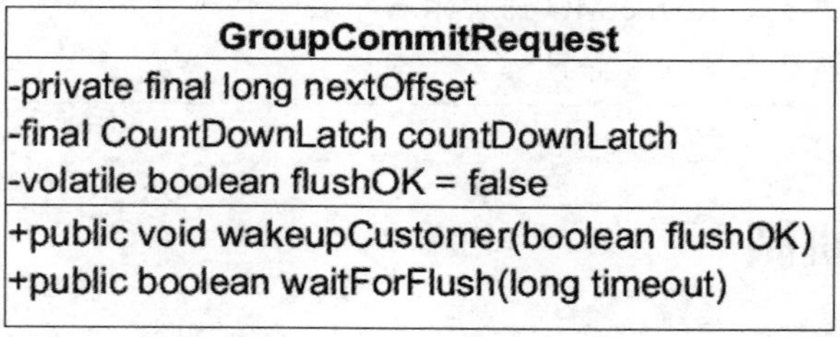

GroupCommitRequest
-private final long nextOffset
-final CountDownLatch countDownLatch
-volatile boolean flushOK = false
+public void wakeupCustomer(boolean flushOK)
+public boolean waitForFlush(long timeout)

图 4-18　GroupCommitRequest 类图

下面让我们一一介绍 GroupCommitRequest 的核心属性。

1）long nextOffset：刷盘点偏移量。

2）CountDownLatch countDownLatch：倒记数锁存器。

3）flushOk：刷盘结果，初始为 false。

代码清单 4-74　GroupCommitRequest#waitForFlush

```
public boolean waitForFlush(long timeout) {
    try {
        this.countDownLatch.await(timeout, TimeUnit.MILLISECONDS);
        return this.flushOK;
```

```
            } catch (InterruptedException e) {
                log.error("Interrupted", e);
                return false;
            }
        }
```

消费发送线程将消息追加到内存映射文件后,将同步任务 GroupCommitRequest 提交到 GroupCommitService 线程,然后调用阻塞等待刷盘结果,超时时间默认为 5s。

代码清单 4-75 GroupCommitRequest#wakeupCustomer

```
public void wakeupCustomer(final boolean flushOK) {
    this.flushOK = flushOK;
    this.countDownLatch.countDown();
}
```

GroupCommitService 线程处理 GroupCommitRequest 对象后将调用 wakeupCustomer 方法将消费发送线程唤醒,并将刷盘告知 GroupCommitRequest。

同步刷盘线程实现 GroupCommitService 类图如图 4-19 所示。

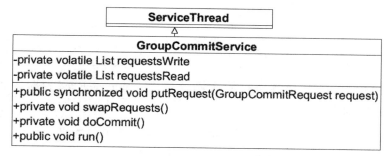

图 4-19 GroupCommitRequest 类图

1) private volatile List<GroupCommitRequest> requestsWrite: 同步刷盘任务暂存容器。

2) private volatile List<GroupCommitRequest> requestsRead: GroupCommitService 线程每次处理的 request 容器,这是一个设计亮点,避免了任务提交与任务执行的锁冲突。

代码清单 4-76 GroupCommitService#putRequest

```
public synchronized void putRequest(final GroupCommitRequest request) {
    synchronized (this.requestsWrite) {
        this.requestsWrite.add(request);
    }
    if (hasNotified.compareAndSet(false, true)) {
        waitPoint.countDown(); // notify
    }
}
```

客户端提交同步刷盘任务到 GroupCommitService 线程,如果该线程处于等待状态则将其唤醒。

代码清单 4-77　GroupCommitService#swapRequests

```
private void swapRequests() {
    List<GroupCommitRequest> tmp = this.requestsWrite;
    this.requestsWrite = this.requestsRead;
    this.requestsRead = tmp;
}
```

由于避免同步刷盘消费任务与其他消息生产者提交任务直接的锁竞争，GroupCommit-Service 提供读容器与写容器，这两个容器每执行完一次任务后，交互，继续消费任务。

代码清单 4-78　GroupCommitService#run

```
public void run() {
    CommitLog.log.info(this.getServiceName() + " service started");
    while (!this.isStopped()) {
        try {
            this.waitForRunning(10);
            this.doCommit();
        } catch (Exception e) {
            CommitLog.log.warn(this.getServiceName() + " service has exception. ", e);
        }
    }
}
```

GroupCommitService 每处理一批同步刷盘请求（requestsRead 容器中请求）后"休息"10ms，然后继续处理下一批，其任务的核心实现为 doCommit 方法。

代码清单 4-79　GroupCommitService#doCommit

```
for (GroupCommitRequest req : this.requestsRead) {
    boolean flushOK = false;
    for (int i = 0; i < 2 && !flushOK; i++) {
        flushOK = CommitLog.this.mappedFileQueue.getFlushedWhere() >=
            req.getNextOffset();
        if (!flushOK) {
                CommitLog.this.mappedFileQueue.flush(0);
        }
    }
    req.wakeupCustomer(flushOK);
}
long storeTimestamp = CommitLog.this.mappedFileQueue.getStoreTimestamp();
if (storeTimestamp > 0) {
        CommitLog.this.defaultMessageStore.getStoreCheckpoint().
            setPhysicMsgTimestamp(storeTimestamp);
}
```

执行刷盘操作，即调用 MappedByteBuffer#force 方法。

1）遍历同步刷盘任务列表，根据加入顺序逐一执行刷盘逻辑。

2）调用 mappedFileQueeu#flush 方法执行刷盘操作，最终会调用 MappedByteBuffer#force() 方法，其具体实现已在 4.4 节中做了详细说明。如果已刷盘指针大于等于提交的刷盘点，表示刷盘成功，每执行一次刷盘操作后，立即调用 GroupCommitRequest#wakeupCustomer 唤醒消息发送线程并通知刷盘结果。

3）处理完所有同步刷盘任务后，更新刷盘检测点 StoreCheckpoint 中的 physicMsgTimestamp，但并没有执行检测点的刷盘操作，刷盘检测点的刷盘操作将在刷写消息队列文件时触发。

同步刷盘的简单描述就是，消息生产者在消息服务端将消息内容追加到内存映射文件中（内存）后，需要同步将内存的内容立刻刷写到磁盘。通过调用内存映射文件（MappedByteBuffer 的 force 方法）可将内存中的数据写入磁盘。

4.8.2 Broker 异步刷盘

代码清单 4-80　CommitLog#handleDiskFlush

```
// Asynchronous flush
else {
if (!this.defaultMessageStore.getMessageStoreConfig().
        isTransientStorePoolEnable()) {
    flushCommitLogService.wakeup();
} else {
    commitLogService.wakeup();
}
```

异步刷盘根据是否开启 transientStorePoolEnable 机制，刷盘实现会有细微差别。如果 transientStorePoolEnable 为 true，RocketMQ 会单独申请一个与目标物理文件（commitlog）同样大小的堆外内存，该堆外内存将使用内存锁定，确保不会被置换到虚拟内存中去，消息首先追加到堆外内存，然后提交到与物理文件的内存映射内存中，再 flush 到磁盘。如果 transientStorePoolEnable 为 flalse，消息直接追加到与物理文件直接映射的内存中，然后刷写到磁盘中。transientStorePoolEnable 为 true 的磁盘刷写流程如图 4-20 所示。

1）首先将消息直接追加到 ByteBuffer（堆外内存 DirectByteBuffer），wrotePosition 随着消息的不断追加向后移动。

2）CommitRealTimeService 线程默认每 200ms 将 ByteBuffer 新追加的内容（wrotePosition 减去 commitedPosition）的数据提交到 MappedByteBuffer 中。

3）MappedByteBuffer 在内存中追加提交的内容，wrotePosition 指针向前后移动，然后返回。

4）commit 操作成功返回，将 commitedPosition 向前后移动本次提交的内容长度，此时 wrotePosition 指针依然可以向前推进。

5）FlushRealTimeService 线程默认每 500ms 将 MappedByteBuffer 中新追加的内存

（wrotePosition 减去上一次刷写位置 flushedPositiont）通过调用 MappedByteBuffer#force() 方法将数据刷写到磁盘。

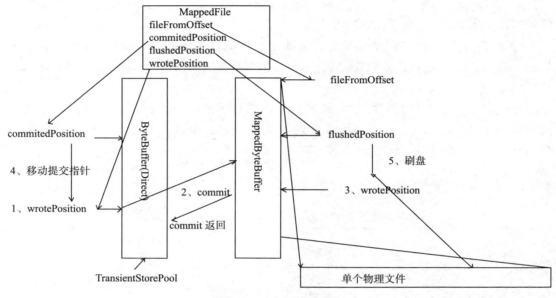

图 4-20　磁盘刷写流程

1. CommitRealTimeService 提交线程工作机制

代码清单 4-81　CommitLog$CommitRealTimeService#run

```
int interval = CommitLog.this.defaultMessageStore.getMessageStoreConfig()
    .getCommitIntervalCommitLog();
int commitDataLeastPages = CommitLog.this.defaultMessageStore.
        getMessageStoreConfig().getCommitCommitLogLeastPages();
int commitDataThoroughInterval = CommitLog.this.defaultMessageStore.
    getMessageStoreConfig().getCommitCommitLogThoroughInterval();
```

Step1：首先解释三个配置参数的含义。

1）commitIntervalCommitLog：CommitRealTimeService 线程间隔时间，默认 200ms。

2）commitCommitLogLeastPages：一次提交任务至少包含页数，如果待提交数据不足，小于该参数配置的值，将忽略本次提交任务，默认 4 页。

3）commitDataThoroughInterval：两次真实提交最大间隔，默认 200ms。

代码清单 4-82　CommitLog$CommitRealTimeService#run

```
long begin = System.currentTimeMillis();
if (begin >= (this.lastCommitTimestamp + commitDataThoroughInterval)) {
    this.lastCommitTimestamp = begin;
    commitDataLeastPages = 0;
}
```

Step2：如果距上次提交间隔超过 commitDataThoroughInterval，则本次提交忽略 commit-CommitLogLeastPages 参数，也就是如果待提交数据小于指定页数，也执行提交操作。

代码清单 4-83　CommitLog$CommitRealTimeService#run

```
boolean result = CommitLog.this.mappedFileQueue.commit(commitDataLeastPages);
long end = System.currentTimeMillis();
if (!result) {
    this.lastCommitTimestamp = end; // result = false means some data committed.
    //now wake up flush thread.
    flushCommitLogService.wakeup();
}
this.waitForRunning(interval);
```

Step3：执行提交操作，将待提交数据提交到物理文件的内存映射内存区，如果返回 false，并不是代表提交失败，而是只提交了一部分数据，唤醒刷盘线程执行刷盘操作。该线程每完成一次提交动作，将等待 200ms 再继续执行下一次提交任务。

2. FlushRealTimeService 刷盘线程工作机制

代码清单 4-84　CommitLog$FlushRealTimeService#run

```
boolean flushCommitLogTimed = CommitLog.this.defaultMessageStore.
        getMessageStoreConfig().isFlushCommitLogTimed();
int interval = CommitLog.this.defaultMessageStore.getMessageStoreConfig().
        getFlushIntervalCommitLog();
int flushPhysicQueueLeastPages = CommitLog.this.defaultMessageStore.
        getMessageStoreConfig().getFlushCommitLogLeastPages();
int flushPhysicQueueThoroughInterval =
        CommitLog.this.defaultMessageStore.getMessageStoreConfig().
        getFlushCommitLogThoroughInterval();
```

Step1：首先解释四个配置参数的含义。

1）flushCommitLogTimed：默认为 false，表示 await 方法等待；如果为 true，表示使用 Thread.sleep 方法等待。

2）flushIntervalCommitLog：FlushRealTimeService 线程任务运行间隔。

3）flushPhysicQueueLeastPages：一次刷写任务至少包含页数，如果待刷写数据不足，小于该参数配置的值，将忽略本次刷写任务，默认 4 页。

4）flushPhysicQueueThoroughInterval：两次真实刷写任务最大间隔，默认 10s。

代码清单 4-85　CommitLog$FlushRealTimeService#run

```
long currentTimeMillis = System.currentTimeMillis();
if (currentTimeMillis >= (this.lastFlushTimestamp +
        flushPhysicQueueThoroughInterval)) {
    this.lastFlushTimestamp = currentTimeMillis;
    flushPhysicQueueLeastPages = 0;
    printFlushProgress = (printTimes++ % 10) == 0;
}
```

Step2：如果距上次提交间隔超过 flushPhysicQueueThoroughInterval，则本次刷盘任务将忽略 flushPhysicQueueLeastPages，也就是如果待刷写数据小于指定页数也执行刷写磁盘操作。

代码清单 4-86　CommitLog$FlushRealTimeService#run

```
if (flushCommitLogTimed) {
    Thread.sleep(interval);
} else {
    this.waitForRunning(interval);
}
```

Step3：执行一次刷盘任务前先等待指定时间间隔，然后再执行刷盘任务。

代码清单 4-87　CommitLog$FlushRealTimeService#run

```
long begin = System.currentTimeMillis();
CommitLog.this.mappedFileQueue.flush(flushPhysicQueueLeastPages);
long storeTimestamp = CommitLog.this.mappedFileQueue.getStoreTimestamp();
if (storeTimestamp > 0) {
    CommitLog.this.defaultMessageStore.getStoreCheckpoint().
        setPhysicMsgTimestamp(storeTimestamp);
}
```

Step4：调用 flush 方法将内存中数据刷写到磁盘，并且更新存储检测点文件的 commitlog 文件的更新时间戳，文件检测点文件（checkpoint 文件）的刷盘动作在刷盘消息消费队列线程中执行，其入口为 DefaultMessageStore#FlushConsumeQueueService。由于消息消费队列、索引文件的刷盘实现原理与 Commitlog 文件的刷盘机制类同，故本书不再做重复分析。

4.9　过期文件删除机制

由于 RocketMQ 操作 CommitLog、ConsumeQueue 文件是基于内存映射机制并在启动的时候会加载 commitlog、ConsumeQueue 目录下的所有文件，为了避免内存与磁盘的浪费，不可能将消息永久存储在消息服务器上，所以需要引入一种机制来删除已过期的文件。RocketMQ 顺序写 Commitlog 文件、ConsumeQueue 文件，所有写操作全部落在最后一个 CommitLog 或 ConsumeQueue 文件上，之前的文件在下一个文件创建后将不会再被更新。RocketMQ 清除过期文件的方法是：如果非当前写文件在一定时间间隔内没有再次被更新，则认为是过期文件，可以被删除，RocketMQ 不会关注这个文件上的消息是否全部被消费。默认每个文件的过期时间为 72 小时，通过在 Broker 配置文件中设置 fileReservedTime 来改变过期时间，单位为小时。接下来详细分析 RocketMQ 是如何设计与实现上述机制的。

代码清单 4-88　DefaultMessageStore#addScheduleTask

```
private void addScheduleTask() {
    this.scheduledExecutorService.scheduleAtFixedRate(new Runnable() {
        public void run() {
            DefaultMessageStore.this.cleanFilesPeriodically();
        }
    }, 1000 * 60, this.messageStoreConfig.getCleanResourceInterval(),
        TimeUnit.MILLISECONDS);
    // ... 省略其他定时任务
}
```

RocketMQ 会每隔 10s 调度一次 cleanFilesPeriodically，检测是否需要清除过期文件。执行频率可以通过设置 cleanResourceInterval，默认为 10s。

代码清单 4-89　DefaultMessageStore#cleanFilesPeriodically

```
private void cleanFilesPeriodically() {
    this.cleanCommitLogService.run();
    this.cleanConsumeQueueService.run();
}
```

分别执行清除消息存储文件（Commitlog 文件）与消息消费队列文件（ConsumeQueue 文件）。由于消息消费队列文件与消息存储文件（Commitlog）共用一套过期文件删除机制，本书将重点讲解消息存储过期文件删除。实现方法：DefaultMessageStore$CleanCommitLogService#deleteExpiredFiles。

代码清单 4-90　DefaultMessageStore$CleanCommitLogService#deleteExpiredFiles

```
long fileReservedTime = DefaultMessageStore.this.getMessageStoreConfig().
        getFileReservedTime();
int deletePhysicFilesInterval = DefaultMessageStore.this.
        getMessageStoreConfig().getDeleteCommitLogFilesInterval();
int destroyMapedFileIntervalForcibly = DefaultMessageStore.this.
        getMessageStoreConfig().getDestroyMapedFileIntervalForcibly();
```

Step1：解释一下这个三个配置属性的含义。

1）fileReservedTime：文件保留时间，也就是从最后一次更新时间到现在，如果超过了该时间，则认为是过期文件，可以被删除。

2）deletePhysicFilesInterval：删除物理文件的间隔，因为在一次清除过程中，可能需要被删除的文件不止一个，该值指定两次删除文件的间隔时间。

3）destroyMapedFileIntervalForcibly：在清除过期文件时，如果该文件被其他线程所占用（引用次数大于 0，比如读取消息），此时会阻止此次删除任务，同时在第一次试图删除该文件时记录当前时间戳，destroyMapedFileIntervalForcibly 表示第一次拒绝删除之后能保留的最大时间，在此时间内，同样可以被拒绝删除，同时会将引用减少 1000 个，超过该时间间隔后，文件将被强制删除。

代码清单4-91 DefaultMessageStore$CleanCommitLogService#deleteExpiredFiles

```
boolean timeup = this.isTimeToDelete();
boolean spacefull = this.isSpaceToDelete();
boolean manualDelete = this.manualDeleteFileSeveralTimes > 0;
if (timeup || spacefull || manualDelete) {
    // 继续执行删除逻辑
    return;
} else {
    // 本次删除任务无作为。
}
```

Step2：RocketMQ 在如下三种情况任意之一满足的情况下将继续执行删除文件操作。

1）指定删除文件的时间点，RocketMQ 通过 deleteWhen 设置一天的固定时间执行一次删除过期文件操作，默认为凌晨 4 点。

2）磁盘空间是否充足，如果磁盘空间不充足，则返回 true，表示应该触发过期文件删除操作。

3）预留，手工触发，可以通过调用 excuteDeleteFilesManualy 方法手工触发过期文件删除，目前 RocketMQ 暂未封装手工触发文件删除的命令。

本节重点分析一下磁盘空间是否充足的实现逻辑。

代码清单4-92 DefaultMessageStore$CleanCommitLogService#isSpaceToDelete

```
private boolean isSpaceToDelete() {
    double ratio = DefaultMessageStore.this.getMessageStoreConfig().
        getDiskMaxUsedSpaceRatio() / 100.0;
    cleanImmediately = false;
    String storePathPhysic = DefaultMessageStore.this.getMessageStoreConfig().
        getStorePathCommitLog();
    double physicRatio=UtilAll.getDiskPartitionSpaceUsedPercent(storePathPhysic);
    if (physicRatio > diskSpaceWarningLevelRatio) {
        boolean diskok = DefaultMessageStore.this.runningFlags.getAndMakeDiskFull();
        // 省略日志输出语句
        cleanImmediately = true;
    } else if (physicRatio > diskSpaceCleanForciblyRatio) {
        cleanImmediately = true;
    } else {
        boolean diskok = DefaultMessageStore.this.runningFlags.getAndMakeDiskOK();
        // 省略日志输出语句
    }
    if (physicRatio < 0 || physicRatio > ratio) {
        return true;
    } // 后面省略对 ConsumeQueue 做同样的判断
    return fasle;
}
```

1）首先解释一下几个参数的含义。

diskMaxUsedSpaceRatio：表示 commitlog、consumequeue 文件所在磁盘分区的最大使

用量，如果超过该值，则需要立即清除过期文件。

cleanImmediately：表示是否需要立即执行清除过期文件操作。

physicRatio：当前 commitlog 目录所在的磁盘分区的磁盘使用率，通过 File#getTotal-Space() 获取文件所在磁盘分区的总容量，通过 File#getFreeSpace() 获取文件所在磁盘分区剩余容量。

diskSpaceWarningLevelRatio：通过系统参数 -Drocketmq.broker.diskSpaceWarningLevelRatio 设置，默认 0.90。如果磁盘分区使用率超过该阈值，将设置磁盘不可写，此时会拒绝新消息的写入。

diskSpaceCleanForciblyRatio：通过系统参数 -Drocketmq.broker.diskSpaceCleanForciblyRatio 设置，默认 0.85。如果磁盘分区使用超过该阈值，建议立即执行过期文件清除，但不会拒绝新消息的写入。

2）如果当前磁盘分区使用率大于 diskSpaceWarningLevelRatio，设置磁盘不可写，应该立即启动过期文件删除操作；如果当前磁盘分区使用率大于 diskSpaceCleanForciblyRatio，建议立即执行过期文件清除；如果磁盘使用率低于 diskSpaceCleanForciblyRatio 将恢复磁盘可写；如果当前磁盘使用率小于 diskMaxUsedSpaceRatio 则返回 false，表示磁盘使用率正常，否则返回 true，需要执行清除过期文件。

代码清单 4-93　MappedFileQueue#deleteExpiredFileByTime

```
for (int i = 0; i < mfsLength; i++) {
    MappedFile mappedFile = (MappedFile) mfs[i];
    long liveMaxTimestamp = mappedFile.getLastModifiedTimestamp() + expiredTime;
    if (System.currentTimeMillis() >= liveMaxTimestamp || cleanImmediately) {
        if (mappedFile.destroy(intervalForcibly)) {
            files.add(mappedFile);
            deleteCount++;
            if (files.size() >= DELETE_FILES_BATCH_MAX) {
                break;
            }
            if (deleteFilesInterval > 0 && (i + 1) < mfsLength) {
                try {
                    Thread.sleep(deleteFilesInterval);
                } catch (InterruptedException e) {
                }
            }
        } else {
            break;
        }
    }
}
```

执行文件销毁与删除。从倒数第二个文件开始遍历，计算文件的最大存活时间（= 文件的最后一次更新时间 + 文件存活时间（默认 72 小时）），如果当前时间大于文件的最大存活时间或需要强制删除文件（当磁盘使用超过设定的阈值）时则执行 MappedFile#destory 方法，清除 MappedFile 占有的相关资源，如果执行成功，将该文件加入到待删除文件列表中，然后统一执行 File#delete 方法将文件从物理磁盘中删除。

4.10 本章小结

RocketMQ 主要存储文件包含消息文件（commitlog）、消息消费队列文件（ConsumeQueue）、Hash 索引文件（IndexFile）、检测点文件（checkpoint）、abort（关闭异常文件）。单个消息存储文件、消息消费队列文件、Hash 索引文件长度固定以便使用内存映射机制进行文件的读写操作。RocketMQ 组织文件以文件的起始偏移量来命名文件，这样根据偏移量能快速定位到真实的物理文件。RocketMQ 基于内存映射文件机制提供了同步刷盘与异步刷盘两种机制，异步刷盘是指在消息存储时先追加到内存映射文件，然后启动专门的刷盘线程定时将内存中的数据刷写到磁盘。

Commitlog，消息存储文件，RocketMQ 为了保证消息发送的高吞吐量，采用单一文件存储所有主题的消息，保证消息存储是完全的顺序写，但这样给文件读取同样带来了不便，为此 RocketMQ 为了方便消息消费构建了消息消费队列文件，基于主题与队列进行组织，同时 RocketMQ 为消息实现了 Hash 索引，可以为消息设置索引键，根据索引能够快速从 Commitog 文件中检索消息。

当消息到达 Commitlog 文件后，会通过 ReputMessageService 线程接近实时地将消息转发给消息消费队列文件与索引文件。为了安全起见，RocketMQ 引入 abort 文件，记录Broker 的停机是正常关闭还是异常关闭，在重启 Broker 时为了保证 Commitlog 文件、消息消费队列文件与 Hash 索引文件的正确性，分别采取不同的策略来恢复文件。

RocketMQ 不会永久存储消息文件、消息消费队列文件，而是启用文件过期机制并在磁盘空间不足或默认在凌晨 4 点删除过期文件，文件默认保存 72 小时并且在删除文件时并不会判断该消息文件上的消息是否被消费。下面一章我们将重点分析有关消息消费的实现机制。

第 5 章 RocketMQ 消息消费

消息成功发送到消息服务器后,接下来需要考虑的问题是如何消费消息,如何整合业务逻辑的处理。本章主要分析 RocketMQ 如何消费消息,重点剖析消息消费的过程中需要解决的问题。

- 消息队列负载与重新分布
- 消息消费模式
- 消息拉取方式
- 消息进度反馈
- 消息过滤
- 顺序消息

5.1 RocketMQ 消息消费概述

消息消费以组的模式开展,一个消费组内可以包含多个消费者,每一个消费组可订阅多个主题,消费组之间有集群模式与广播模式两种消费模式。集群模式,主题下的同一条消息只允许被其中一个消费者消费。广播模式,主题下的同一条消息将被集群内的所有消费者消费一次。消息服务器与消费者之间的消息传送也有两种方式:推模式、拉模式。所谓的拉模式,是消费端主动发起拉消息请求,而推模式是消息到达消息服务器后,推送给消息消费者。RocketMQ 消息推模式的实现基于拉模式,在拉模式上包装一层,一个拉取任务完成后开始下一个拉取任务。

集群模式下,多个消费者如何对消息队列进行负载呢?消息队列负载机制遵循一个通用的思想：一个消息队列同一时间只允许被一个消费者消费，一个消费者可以消费多个消息队列。

RocketMQ 支持局部顺序消息消费，也就是保证同一个消息队列上的消息顺序消费。不支持消息全局顺序消费，如果要实现某一主题的全局顺序消息消费，可以将该主题的队列数设置为 1，牺牲高可用性。

RocketMQ 支持两种消息过滤模式：表达式（TAG、SQL92）与类过滤模式。

消息拉模式，主要是由客户端手动调用消息拉取 API，而消息推模式是消息服务器主动将消息推送到消息消费端，本章将以推模式为突破口重点介绍 RocketMQ 消息消费实现原理。

5.2 消息消费者初探

消息消费分为推和拉两种模式，下面我们介绍推模式的消费者 MQPushConsume 的主要 API，如图 5-1 所示。

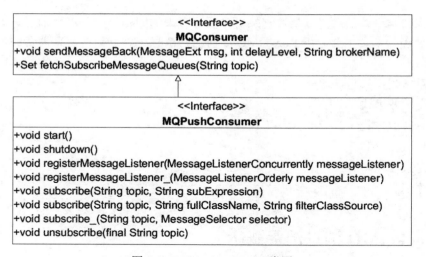

图 5-1　MQPushConsumer 类图

下面让我们来一一介绍 MQPushConsumer 的核心属性。

1）void sendMessageBack（MessageExt msg, int delayLevel, String brokerName）
　　发送消息 ACK 确认。
　　msg：消息。
　　delayLevel：消息延迟级别。
　　broderName：消息服务器名称。

2）Set<MessageQueue> fetchSubscribeMessageQueues（final String topic）

获取消费者对主题 topic 分配了哪些消息队列。

topic：主题名称。

3）void registerMessageListener（final MessageListenerConcurrently messageListener）
注册并发消息事件监听器。

4）void registerMessageListener（final MessageListenerOrderly messageListener）
注册顺序消费事件监听器。

5）void subscribe（final String topic, final String subExpression）
基于主题订阅消息。

topic：消息主题。

subExpression：消息过滤表达式，TAG 或 SQL92 表达式。

6）void subscribe（final String topic, final String fullClassName,final String filterClassSource）
基于主题订阅消息，消息过滤方式使用类模式。

topic：消息主题。

fullClassName：过滤类全路径名。

filterClassSource：过滤类代码。

7）void unsubscribe（final String topic）
取消消息订阅。

DefaultMQPushConsumer（推模式消息消费者）主要属性如图 5-2 所示。

```
DefaultMQPushConsumer
-protected DefaultMQPushConsumerImpl defaultMQPushConsumerImpl
-private String consumerGroup
-private MessageModel messageModel = MessageModel.CLUSTERING
-private ConsumeFromWhere consumeFromWhere
-private AllocateMessageQueueStrategy allocateMessageQueueStrategy
-private Map subscription
-private MessageListener messageListener
-private OffsetStore offsetStore
-private int consumeThreadMin = 20
-private int consumeThreadMax = 64
-private int consumeConcurrentlyMaxSpan = 2000
-private int pullThresholdForQueue = 1000
-private long pullInterval = 0
-private int consumeMessageBatchMaxSize = 1
-private int pullBatchSize = 32
-private boolean postSubscriptionWhenPull = false
-private int maxReconsumeTimes = -1
-private long suspendCurrentQueueTimeMillis = 1000
-private long consumeTimeout = 15
```

图 5-2 DefaultMQPushConsumer 类图

1）consumerGroup：消费者所属组。

2）messageModel：消息消费模式，分为集群模式、广播模式，默认为集群模式。

3）ConsumeFromWhere consumeFromWhere，根据消息进度从消息服务器拉取不到消息时重新计算消费策略。

CONSUME_FROM_LAST_OFFSET：从队列当前最大偏移量开始消费。

CONSUME_FROM_FIRST_OFFSET：从队列当前最小偏移量开始消费。

CONSUME_FROM_TIMESTAMP：从消费者启动时间戳开始消费。

注意：如果从消息进度服务 OffsetStore 读取到 MessageQueue 中的偏移量不小于 0，则使用读取到的偏移量，只有在读到的偏移量小于 0 时，上述策略才会生效。

4）allocateMessageQueueStrategy：集群模式下消息队列负载策略。

5）Map<String /* topic */, String /* sub expression */> subscription：订阅信息。

6）MessageListener messageListener：消息业务监听器。

7）private OffsetStore offsetStore：消息消费进度存储器。

8）int consumeThreadMin = 20，消息者最新线程数。

9）int consumeThreadMax = 64，消费者最大线程数，由于消费者线程池使用无界队列，故消费者线程个数其实最多只有 consumeThreadMin 个。

10）consumeConcurrentlyMaxSpan，并发消息消费时处理队列最大跨度，默认 2000，表示如果消息处理队列中偏移量最大的消息与偏移量最小的消息的跨度超过 2000 则延迟 50 毫秒后再拉取消息。

11）int pullThresholdForQueue：默认值 1000，每 1000 次流控后打印流控日志。

12）long pullInterval = 0，推模式下拉取任务间隔时间，默认一次拉取任务完成继续拉取。

13）int pullBatchSize：每次消息拉取所拉取的条数，默认 32 条。

14）int consumeMessageBatchMaxSize：消息并发消费时一次消费消息条数，通俗点说就是每次传入 MessageListtener#consumeMessage 中的消息条数。

15）postSubscriptionWhenPull：是否每次拉取消息都更新订阅信息，默认为 false。

16）maxReconsumeTimes：最大消费重试次数。如果消息消费次数超过 maxReconsumeTimes 还未成功，则将该消息转移到一个失败队列，等待被删除。

17）suspendCurrentQueueTimeMillis：延迟将该队列的消息提交到消费者线程的等待时间，默认延迟 1s。

18）long consumeTimeout，消息消费超时时间，默认为 15，单位为分钟。

5.3 消费者启动流程

消息消费者是如何启动的，请跟我一起来分析 DefaultMQPushConsumerImpl 的 start 方法，具体代码如下。

代码清单 5-1　DefaultMQPushConsumerImpl#copySubscription

```java
private void copySubscription() throws MQClientException {
    try {
        Map<String, String> sub = this.defaultMQPushConsumer.getSubscription();
        if (sub != null) {
            for (final Map.Entry<String, String> entry : sub.entrySet()) {
                final String topic = entry.getKey();
                final String subString = entry.getValue();
                SubscriptionData subscriptionData = FilterAPI.buildSubscription
                    Data(this.defaultMQPushConsumer.getConsumerGroup(),
                        topic, subString);
                this.rebalanceImpl.getSubscriptionInner().put(topic,
                    subscriptionData);
            }
        }
        if (null == this.messageListenerInner) {
            this.messageListenerInner =
                this.defaultMQPushConsumer.getMessageListener();
        }
        switch (this.defaultMQPushConsumer.getMessageModel()) {
            case BROADCASTING:
                break;
            case CLUSTERING:
                final String retryTopic =
                    MixAll.getRetryTopic(this.defaultMQPushConsumer.getConsumer
                        Group());
                SubscriptionData subscriptionData = FilterAPI.
                    buildSubscriptionData(this.defaultMQPushConsumer.getConsume
                        rGroup(),retryTopic, SubscriptionData.SUB_ALL);
                this.rebalanceImpl.getSubscriptionInner().put(retryTopic,
                    subscriptionData);
                break;
            default:
                break;
        }
    } catch (Exception e) {
        throw new MQClientException("subscription exception", e);
    }
}
```

Step1：构建主题订阅信息 SubscriptionData 并加入到 RebalanceImpl 的订阅消息中。订阅关系来源主要有两个。

1）通过调用 DefaultMQPushConsumerImpl#subscribe（String topic, String subExpression）方法。

2）订阅重试主题消息。从这里可以看出，RocketMQ 消息重试是以消费组为单位，而不是主题，消息重试主题名为 %RETRY%+ 消费组名。消费者在启动的时候会自动订阅该主题，参与该主题的消息队列负载。

代码清单 5-2　DefaultMQPushConsumerImpl#start

```
if (this.defaultMQPushConsumer.getMessageModel() == MessageModel.CLUSTERING) {
    this.defaultMQPushConsumer.changeInstanceNameToPID();
}
this.mQClientFactory = MQClientManager.getInstance().
        getAndCreateMQClientInstance(this.defaultMQPushConsumer, this.rpcHook);
this.rebalanceImpl.setConsumerGroup(this.defaultMQPushConsumer.
        getConsumerGroup());
this.rebalanceImpl.setMessageModel(this.defaultMQPushConsumer.
        getMessageModel());
this.rebalanceImpl.setAllocateMessageQueueStrategy(
this.defaultMQPushConsumer.getAllocateMessageQueueStrategy());
this.rebalanceImpl.setmQClientFactory(this.mQClientFactory);
this.pullAPIWrapper = new PullAPIWrapper(mQClientFactory,
this.defaultMQPushConsumer.getConsumerGroup(), isUnitMode());
this.pullAPIWrapper.registerFilterMessageHook(filterMessageHookList);
```

Step2：初始化 MQClientInstance、RebalanceImple（消息重新负载实现类）等。

代码清单 5-3　DefaultMQPushConsumerImpl#start

```
if (this.defaultMQPushConsumer.getOffsetStore() != null) {
    this.offsetStore = this.defaultMQPushConsumer.getOffsetStore();
} else {
    switch (this.defaultMQPushConsumer.getMessageModel()) {
    case BROADCASTING:
        this.offsetStore = new LocalFileOffsetStore(
            this.mQClientFactory,this.defaultMQPushConsumer.getConsumerGroup());
        break;
    case CLUSTERING:
        this.offsetStore = new RemoteBrokerOffsetStore(this.mQClientFactory,
            this.defaultMQPushConsumer.getConsumerGroup());
        break;
    default:
        break;
    }
}
this.offsetStore.load();
```

Step3：初始化消息进度。如果消息消费是集群模式，那么消息进度保存在 Broker 上；如果是广播模式，那么消息消费进度存储在消费端。具体实现细节后面将重点探讨。

代码清单 5-4　DefaultMQPushConsumerImpl#start

```
if (this.getMessageListenerInner() instanceof MessageListenerOrderly) {
    this.consumeOrderly = true;
    this.consumeMessageService = new ConsumeMessageOrderlyService(this,
        (MessageListenerOrderly) this.getMessageListenerInner());
} else if (this.getMessageListenerInner() instanceof
        MessageListenerConcurrently) {
    this.consumeOrderly = false;
```

```
            this.consumeMessageService = new ConsumeMessageConcurrentlyService(this,
                (MessageListenerConcurrently) this.getMessageListenerInner());
    }
    this.consumeMessageService.start();
```

Step4：根据是否是顺序消费，创建消费端消费线程服务。ConsumeMessageService 主要负责消息消费，内部维护一个线程池。

代码清单 5-5　DefaultMQPushConsumerImpl#start

```
boolean registerOK = mQClientFactory.registerConsumer(
            this.defaultMQPushConsumer.getConsumerGroup(), this);
if (!registerOK) {
    this.serviceState = ServiceState.CREATE_JUST;
    this.consumeMessageService.shutdown();
    throw new MQClientException("The consumer group[" +
        this.defaultMQPushConsumer.getConsumerGroup()
            + "] has been created before, specify another name please."
        + FAQUrl.suggestTodo(FAQUrl.GROUP_NAME_DUPLICATE_URL),null);
}
mQClientFactory.start();
```

Step5：向 MQClientInstance 注册消费者，并启动 MQClientInstance，在一个 JVM 中的所有消费者、生产者持有同一个 MQClientInstance，MQClientInstance 只会启动一次。

5.4　消息拉取

本节将基于 PUSH 模式来详细分析消息拉取机制。消息消费有两种模式：广播模式与集群模式，广播模式比较简单，每一个消费者需要去拉取订阅主题下所有消费队列的消息，本节主要基于集群模式。在集群模式下，同一个消费组内有多个消息消费者，同一个主题存在多个消费队列，那么消费者如何进行消息队列负载呢？从上文启动流程也知道，每一个消费组内维护一个线程池来消费消息，那么这些线程又是如何分工合作的呢？

消息队列负载，通常的做法是一个消息队列在同一时间只允许被一个消息消费者消费，一个消息消费者可以同时消费多个消息队列，那么 RocketMQ 是如何实现的呢？带着上述问题，我们开始 RocketMQ 消息消费机制的探讨。

从 MQClientInstance 的启动流程中可以看出，RocketMQ 使用一个单独的线程 PullMessageService 来负责消息的拉取。

5.4.1　PullMessageService 实现机制

PullMessageService 继承的是 ServiceThread，从名称来看，它是服务线程，通过 run 方法启动，具体代码如下。

代码清单 5-6 PullMessageService#run

```
public void run() {
    log.info(this.getServiceName() + " service started");
    while (!this.isStopped()) {
        try {
            PullRequest pullRequest = this.pullRequestQueue.take();
            if (pullRequest != null) {
                this.pullMessage(pullRequest);
            }
        } catch (InterruptedException e) {
        } catch (Exception e) {
            log.error("Pull Message Service Run Method exception", e);
        }
    }
    log.info(this.getServiceName() + " service end");
}
```

PullMessageService，消息拉取服务线程，run 方法是其核心逻辑。run 方法的几个核心要点如下。

1）while（!this.isStopped()）这是一种通用的设计技巧，stopped 声明为 volatile，每执行一次业务逻辑检测一下其运行状态，可以通过其他线程将 stopped 设置为 true 从而停止该线程。

2）从 pullRequestQueue 中获取一个 PullRequest 消息拉取任务，如果 pullRequestQueue 为空，则线程将阻塞，直到有拉取任务被放入。

3）调用 pullMessage 方法进行消息拉取。

那 PullRequest 是什么时候添加的呢？

代码清单 5-7 PullMessageService#executePullRequestLater executePullRequestImmediately

```
public void executePullRequestLater(final PullRequest pullRequest, final long
        timeDelay) {
    this.scheduledExecutorService.schedule(new Runnable() {
        public void run() {
            PullMessageService.this.executePullRequestImmediately(pullRequest);
        }
    }, timeDelay, TimeUnit.MILLISECONDS);
}
public void executePullRequestImmediately(final PullRequest pullRequest) {
    try {
        this.pullRequestQueue.put(pullRequest);
    } catch (InterruptedException e) {
        log.error("executePullRequestImmediately pullRequestQueue.put", e);
    }
}
```

原来，PullMessageService 提供延迟添加与立即添加 2 种方式将 PullRequest 放入到

pullRequestQueue 中。那 PullRequest 在什么时候创建呢？executePullRequestImmediately 方法调用链如图 5-3 所示。

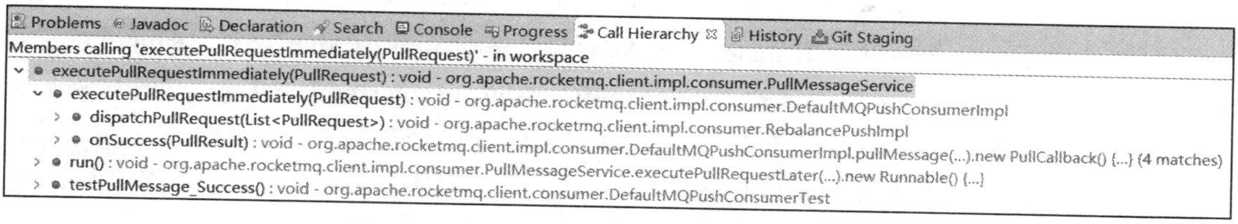

图 5-3 executePullRequestImmediately 调用链

通过跟踪发现，主要有两个地方会调用，一个是在 RocketMQ 根据 PullRequest 拉取任务执行完一次消息拉取任务后，又将 PullRequest 对象放入到 pullRequestQueue，第二个是在 RebalancceImpl 中创建。RebalanceImpl 就是下节重点要介绍的消息队列负载机制，也就是 PullRequest 对象真正创建的地方。

从上面分析可知，PullMessageService 只有在拿到 PullRequest 对象时才会执行拉取任务，那么 PullRequest 究竟是什么呢？其类图如图 5-4 所示。

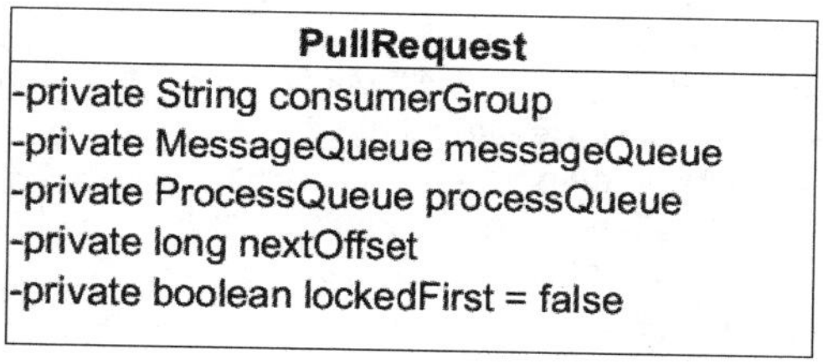

```
PullRequest
-private String consumerGroup
-private MessageQueue messageQueue
-private ProcessQueue processQueue
-private long nextOffset
-private boolean lockedFirst = false
```

图 5-4 PullRequest 类图

下面让我们一一来介绍一下 PullRequest 的核心属性。

1）String consumerGroup：消费者组。

2）MessageQueue messageQueue：待拉取消费队列。

3）ProcessQueue processQueue：消息处理队列，从 Broker 拉取到的消息先存入 ProccessQueue，然后再提交到消费者消费线程池消费。

4）long nextOffset：待拉取的 MessageQueue 偏移量。

5）Boolean lockedFirst：是否被锁定。

代码清单 5-8 PullMessageService#pullMessage

```
private void pullMessage(final PullRequest pullRequest) {
    final MQConsumerInner consumer =
        this.mQClientFactory.selectConsumer(pullRequest.getConsumerGroup());
    if (consumer != null) {
        DefaultMQPushConsumerImpl impl = (DefaultMQPushConsumerImpl) consumer;
        impl.pullMessage(pullRequest);
    } else {
```

```
            log.warn("No matched consumer for the PullRequest {}, drop it", pullRequest);
        }
    }
```

根据消费组名从 MQClientInstance 中获取消费者内部实现类 MQConsumerInner，令人意外的是这里将 consumer 强制转换为 DefaultMQPushConsumerImpl，也就是 PullMessage-Service，该线程只为 PUSH 模式服务，那拉模式如何拉取消息呢？其实细想也不难理解，PULL 模式，RocketMQ 只需要提供拉取消息 API 即可，具体由应用程序显示调用拉取 API。

5.4.2 ProcessQueue 实现机制

ProcessQueue 是 MessageQueue 在消费端的重现、快照。PullMessageService 从消息服务器默认每次拉取 32 条消息，按消息的队列偏移量顺序存放在 ProcessQueue 中，PullMessageService 然后将消息提交到消费者消费线程池，消息成功消费后从 ProcessQueue 中移除。ProcessQueue 的类图如图 5-5 所示。

```
                        ProcessQueue
-ReadWriteLock lockTreeMap
-TreeMap msgTreeMap
-AtomicLong msgCount = new AtomicLong()
-Lock lockConsume = new ReentrantLock()
-TreeMap msgTreeMapTemp
-AtomicLong tryUnlockTimes = new AtomicLong(0)
-volatile long queueOffsetMax = 0L
-volatile boolean dropped = false
-volatile long lastPullTimestamp
-volatile long lastConsumeTimestamp
-volatile boolean locked = false
-volatile long lastLockTimestamp
-volatile boolean consuming = false
-volatile long msgAccCnt = 0
+public boolean isLockExpired()
+public boolean isPullExpired()
+public void cleanExpiredMsg(DefaultMQPushConsumer pushConsumer)
+public boolean putMessage(final List msgs)
+public long getMaxSpan()
+public long removeMessage(final List msgs)
+public void rollback()
+public long commit()
+public void makeMessageToCosumeAgain(List msgs)
+public List takeMessags(final int batchSize)
```

图 5-5　PocessQueue 类图

1. ProccessQueue 核心属性

1）TreeMap<Long, MessageExt> msgTreeMap：消息存储容器，键为消息在 ConsumeQueue

中的偏移量，MessageExt：消息实体。

2）TreeMap<Long, MessageExt> msgTreeMapTemp：消息临时存储容器，键为消息在 ConsumeQueue 中的偏移量，MessageExt 为消息实体，该结构用于处理顺序消息，消息消费线程从 ProcessQueue 的 msgTreeMap 中取出消息前，先将消息临时存储在 msgTreeMap-Temp 中。

3）ReadWriteLock lockTreeMap = new ReentrantReadWriteLock()：读写锁，控制多线程并发修改 msgTreeMap、msgTreeMapTemp。

4）AtomicLong msgCount：ProcessQueue 中总消息数。

5）volatile long queueOffsetMax：当前 ProcessQueue 中包含的最大队列偏移量。

6）volatile boolean dropped = false：当前 ProccesQueue 是否被丢弃。

7）volatile long lastPullTimestamp：上一次开始消息拉取时间戳。

8）volatile long lastConsumeTimestamp：上一次消息消费时间戳。

2. ProcessQueue 核心方法

1）public boolean isLockExpired()

判断锁是否过期，锁超时时间默认为 30s。可以通过系统参数 rocketmq.client.rebalance.lockMaxLiveTime 来设置。

2）public boolean isPullExpired()

判断 PullMessageService 是否空闲，默认 120s，通过系统参数 rocketmq.client.pull.pullMaxIdleTime 来设置。

3）public void cleanExpiredMsg（DefaultMQPushConsumer pushConsumer）

移除消费超时的消息，默认超过 15 分钟未消费的消息将延迟 3 个延迟级别再消费。

4）public boolean putMessage（final List<MessageExt> msgs）

添加消息，PullMessageService 拉取消息后，先调用该方法将消息添加到 ProcessQueue。

5）public long getMaxSpan()

获取当前消息最大间隔。getMaxSpan() / 20 并不能说明 Procequeue 中包含的消息个数，但是能说明当前处理队列中第一条消息与最后一条消息的偏移量已经超过的消息个数。

6）public boolean isLockExpired()

7）public long removeMessage（final List<MessageExt> msgs）

移除消息。

8）public void rollback()

将 msgTreeMapTmp 中所有消息重新放入到 msgTreeMap 并清除 msgTreeMap-Tmp。

9）public long commit()

  将 msgTreeMapTmp 中的消息清除，表示成功处理该批消息。

10）public void makeMessageToCosumeAgain（List<MessageExt> msgs）

  重新消费该批消息。

11）public List<MessageExt> takeMessags（final int batchSize）

  从 ProcessQueue 中取出 batchSize 条消息。

5.4.3 消息拉取基本流程

本节将以并发消息消费来探讨整个消息消费流程，顺序消息的实现原理将在 5.9 节中单独分析。

消息拉取分为 3 个主要步骤。

1）消息拉取客户端消息拉取请求封装。

2）消息服务器查找并返回消息。

3）消息拉取客户端处理返回的消息。

1. 客户端封装消息拉取请求

消息拉取入口：DefaultMQPushConsumerImpl#pullMessage。

代码清单 5-9　DefaultMQPushConsumerImpl#pullMessage

```
final ProcessQueue processQueue = pullRequest.getProcessQueue();
if (processQueue.isDropped()) {
    log.info("the pull request[{}] is dropped.", pullRequest.toString());
    return;
}
pullRequest.getProcessQueue().setLastPullTimestamp(
        System.currentTimeMillis());
try {
    this.makeSureStateOK();
} catch (MQClientException e) {
    log.warn("pullMessage exception, consumer state not ok", e);
    this.executePullRequestLater(pullRequest,
            PULL_TIME_DELAY_MILLS_WHEN_EXCEPTION);
    return;
}
if (this.isPause()) {
    this.executePullRequestLater(pullRequest,
            PULL_TIME_DELAY_MILLS_WHEN_SUSPEND);
    return;
}
```

Step1：从 PullRequest 中获取 ProcessQueue，如果处理队列当前状态未被丢弃，则更新 ProcessQueue 的 lastPullTimestamp 为当前时间戳；如果当前消费者被挂起，则将拉取任务延迟 1s 再次放入到 PullMessageService 的拉取任务队列中，结束本次消息拉取。

代码清单 5-10　DefaultMQPushConsumerImpl#pullMessage

```
long size = processQueue.getMsgCount().get();
if (size > this.defaultMQPushConsumer.getPullThresholdForQueue()) {
    this.executePullRequestLater(pullRequest,PULL_TIME_DELAY_MILLS_WHEN_FLOW_C
        ONTROL);
    if ((flowControlTimes1++ % 1000) == 0) {
        // 省略流控输出语句
    }
    return;
}
if (processQueue.getMaxSpan() >
    this.defaultMQPushConsumer.getConsumeConcurrentlyMaxSpan()) {
    this.executePullRequestLater(pullRequest,
        PULL_TIME_DELAY_MILLS_WHEN_FLOW_CONTROL);
    if ((flowControlTimes2++ % 1000) == 0) {
        // 省略流控输出语句
    }
    return;
}
```

Step2：进行消息拉取流控。从消息消费数量与消费间隔两个维度进行控制。

1）消息处理总数，如果 ProcessQueue 当前处理的消息条数超过了 pullThresholdForQueue=1000 将触发流控，放弃本次拉取任务，并且该队列的下一次拉取任务将在 50 毫秒后才加入到拉取任务队列中，每触发 1000 次流控后输出提示语：the consumer message buffer is full, so do flow control, minOffset={ 队列最小偏移量 }, maxOffset={ 队列最大偏移量 }, size={ 消息总条数 }, pullRequest={ 拉取任务 }, flowControlTimes={ 流控触发次数 }。

2）ProcessQueue 中队列最大偏移量与最小偏离量的间距，不能超过 consumeConcurrentlyMaxSpan，否则触发流控，每触发 1000 次输出提示语：the queue's messages, span too long, so do flow control, minOffset={ 队列最小偏移量 }, maxOffset={ 队列最大偏移量 }, maxSpan={ 间隔 }, pullRequest={ 拉取任务信息 }, flowControlTimes={ 流控触发次数 }。这里主要的考量是担心一条消息堵塞，消息进度无法向前推进，可能造成大量消息重复消费。

代码清单 5-11　DefaultMQPushConsumerImpl#pullMessage

```
final SubscriptionData subscriptionData = this.rebalanceImpl.
    getSubscriptionInner().get(pullRequest.getMessageQueue().getTopic());
if (null == subscriptionData) {
    this.executePullRequestLater(pullRequest,PULL_TIME_DELAY_MILLS_WHEN_EXCEPTI
        ON);
    return;
}
```

Step3：拉取该主题订阅信息，如果为空，结束本次消息拉取，关于该队列的下一次拉取任务延迟 3s。

代码清单 5-12 DefaultMQPushConsumerImpl#pullMessage

```
boolean commitOffsetEnable = false;
long commitOffsetValue = 0L;
if (MessageModel.CLUSTERING == this.defaultMQPushConsumer.getMessageModel()) {
    commitOffsetValue =
        this.offsetStore.readOffset(pullRequest.getMessageQueue(),
        ReadOffsetType.READ_FROM_MEMORY);
    if (commitOffsetValue > 0) {
        commitOffsetEnable = true;
    }
}
String subExpression = null;
boolean classFilter = false;
if (sd != null) {
    if (this.defaultMQPushConsumer.isPostSubscriptionWhenPull()
        && !sd.isClassFilterMode()) {
        subExpression = sd.getSubString();
    }
    classFilter = sd.isClassFilterMode();
}
int sysFlag = PullSysFlag.buildSysFlag(commitOffsetEnable, // commitOffset
    true, subExpression != null, classFilter // class filter);
```

Step4：构建消息拉取系统标记，拉消息系统标记如图 5-6 所示。

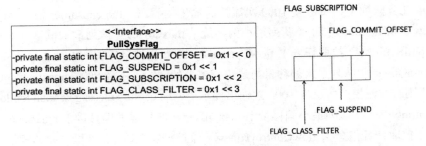

图 5-6 PullSysFlag 类图

下面让我们一一来介绍 PullSysFlag 的枚举值含义。

FLAG_COMMIT_OFFSET：表示从内存中读取的消费进度大于 0，则设置该标记位。

FLAG_SUSPEND：表示消息拉取时支持挂起。

FLAG_SUBSCRIPTION：消息过滤机制为表达式，则设置该标记位。

FLAG_CLASS_FILTER：消息过滤机制为类过滤模式。

代码清单 5-13 DefaultMQPushConsumerImpl#pullMessage

```
this.pullAPIWrapper.pullKernelImpl(pullRequest.getMessageQueue(),
    subExpression,subscriptionData.getExpressionType(),
```

```
            subscriptionData.getSubVersion(),pullRequest.getNextOffset(),
            this.defaultMQPushConsumer.getPullBatchSize(),sysFlag,commitOffsetValue,
            BROKER_SUSPEND_MAX_TIME_MILLIS,CONSUMER_TIMEOUT_MILLIS_WHEN_SUSPEND,
            CommunicationMode.ASYNC,pullCallback);
```

Step5：调用 PullAPIWrapper.pullKernelImpl 方法后与服务端交互，调用 pullKernelImpl 方法之前先了解一下其参数含义。

1）MessageQueue mq：从哪个消息消费队列拉取消息。

2）String subExpression：消息过滤表达式。

3）String expressionType：消息表达式类型，分为 TAG、SQL92。

4）long offset：消息拉取偏移量。

5）int maxNums：本次拉取最大消息条数，默认 32 条。

6）int sysFlag：拉取系统标记。

7）long commitOffset：当前 MessageQueue 的消费进度（内存中）。

8）long brokerSuspendMaxTimeMillis：消息拉取过程中允许 Broker 挂起时间，默认 15s。

9）long timeoutMillis：消息拉取超时时间。

10）CommunicationMode communicationMode：消息拉取模式，默认为异步拉取。

11）PullCallback pullCallback：从 Broker 拉取到消息后的回调方法。

代码清单 5-14　PullAPIWrapper#pullKernelImpl

```
FindBrokerResult findBrokerResult =
        this.mQClientFactory.findBrokerAddressInSubscribe(mq.getBrokerName(),
            this.recalculatePullFromWhichNode(mq), false);
if (null == findBrokerResult) {
    this.mQClientFactory.updateTopicRouteInfoFromNameServer(mq.getTopic());
    findBrokerResult = this.mQClientFactory.findBrokerAddressInSubscribe(
        mq.getBrokerName(),this.recalculatePullFromWhichNode(mq), false);
}
```

Step6：根据 brokerName、BrokerId 从 MQClientInstance 中获取 Broker 地址，在整个 RocketMQ Broker 的部署结构中，相同名称的 Broker 构成主从结构，其 BrokerId 会不一样，在每次拉取消息后，会给出一个建议，下次拉取从主节点还是从节点拉取，其类图如图 5-7 所示。

下面让我们一一来介绍 FindBrokerResult 的核心属性。

String brokerAddr：Broker 地址。

bollean slave：是否是从节点。

int brokerVersion：Broker 版本。

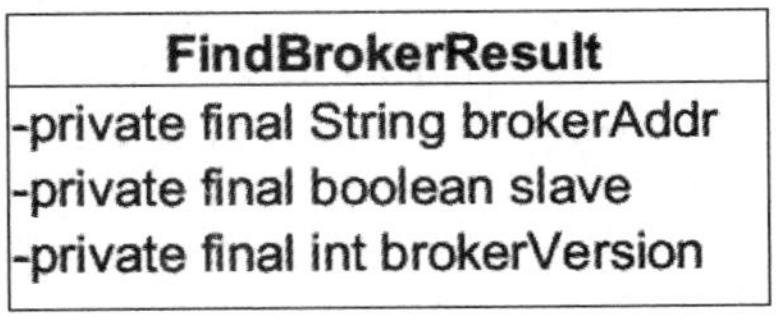

FindBrokerResult
-private final String brokerAddr
-private final boolean slave
-private final int brokerVersion

图 5-7　Find BrokerResult 类图

代码清单 5-15　PullAPIWrapper#pullKernelImpl

```
String brokerAddr = findBrokerResult.getBrokerAddr();
if (PullSysFlag.hasClassFilterFlag(sysFlagInner)) {
    brokerAddr = computPullFromWhichFilterServer(mq.getTopic(), brokerAddr);
}
```

Step7：如果消息过滤模式为类过滤，则需要根据主题名称、broker 地址找到注册在 Broker 上的 FilterServer 地址，从 FilterServer 上拉取消息，否则从 Broker 上拉取消息。

上述步骤完成后，RocketMQ 通过 MQClientAPIImpl#pullMessageAsync 方法异步向 Broker 拉取消息。

2. 消息服务端 Broker 组装消息

根据消息拉取命令 Code：RequestCode.PULL_MESSAGE，很容易找到 Brokder 端处理消息拉取的入口：org.apache.rocketmq.broker.processor.PullMessageProcessor#processRequest。

Step1：根据订阅信息，构建消息过滤器。本部分将在 5.8 节、第 6 章详细讲解。

代码清单 5-16　PullMessageProcessor#processRequest

```
final GetMessageResult getMessageResult =this.brokerController.
    getMessageStore().getMessage(requestHeader.getConsumerGroup(),
    requestHeader.getTopic(),requestHeader.getQueueId(),requestHeader.getQueueO
    ffset(), requestHeader.getMaxMsgNums(), messageFilter);
```

Step2：调用 MessageStore.getMessage 查找消息，该方法参数的含义如下。

1）String group：消费组名称。

2）String topic：主题名称。

3）int queueId：队列 ID。

4）long offset：待拉取偏移量。

5）int maxMsgNums：最大拉取消息条数。

6）MessageFilter messageFilter：消息过滤器。

代码清单 5-17　DefaultMessageStore#getMessage

```
GetMessageStatus status = GetMessageStatus.NO_MESSAGE_IN_QUEUE;
long nextBeginOffset = offset;
long minOffset = 0;
long maxOffset = 0;
GetMessageResult getResult = new GetMessageResult();
final long maxOffsetPy = this.commitLog.getMaxOffset();
ConsumeQueue consumeQueue = findConsumeQueue(topic, queueId);
```

Step3：根据主题名称与队列编号获取消息消费队列。

nextBeginOffset：待查找的队列偏移量。

minOffset：当前消息队列最小偏移量。

maxOffset：当前消息队列最大偏移量。

maxOffsetPy：当前 commitlog 文件最大偏移量。

代码清单 5-18　DefaultMessageStore#getMessage
```
minOffset = consumeQueue.getMinOffsetInQueue();
maxOffset = consumeQueue.getMaxOffsetInQueue();
if (maxOffset == 0) {
    status = GetMessageStatus.NO_MESSAGE_IN_QUEUE;
    nextBeginOffset = nextOffsetCorrection(offset, 0);
} else if (offset < minOffset) {
    status = GetMessageStatus.OFFSET_TOO_SMALL;
    nextBeginOffset = nextOffsetCorrection(offset, minOffset);
} else if (offset == maxOffset) {
    status = GetMessageStatus.OFFSET_OVERFLOW_ONE;
    nextBeginOffset = nextOffsetCorrection(offset, offset);
} else if (offset > maxOffset) {
    status = GetMessageStatus.OFFSET_OVERFLOW_BADLY;
    if (0 == minOffset) {
      nextBeginOffset = nextOffsetCorrection(offset, minOffset);
    } else {
      nextBeginOffset = nextOffsetCorrection(offset, maxOffset);
    }
}
```

Step4：消息偏移量异常情况校对下一次拉取偏移量。

1）maxOffset = 0，表示当前消费队列中没有消息，拉取结果：NO_MESSAGE_IN_QUEUE。

如果当前 Broker 为主节点或 offsetCheckInSlave 为 false，下次拉取偏移量依然为 offset。

如果当前 Broker 为从节点，offsetCheckInSlave 为 true，设置下次拉取偏移量为 0。

2）offset < minOffset，表示待拉取消息偏移量小于队列的起始偏移量，拉取结果为：OFFSET_TOO_SMALL。

如果当前 Broker 为主节点或 offsetCheckInSlave 为 false，下次拉取偏移量依然为 offset。

如果当前 Broker 为从节点并且 offsetCheckInSlave 为 true，下次拉取偏移量设置为 minOffset。

3）offset == maxOffset，如果待拉取偏移量等于队列最大偏移量，拉取结果：OFFSET_OVERFLOW_ONE。下次拉取偏移量依然为 offset。

4）Offset > maxOffset，表示偏移量越界，拉取结果：OFFSET_OVERFLOW_BADLY。

根据是否是主节点、从节点，同样校对下次拉取偏移量。

上述异常情况很关键，请留意客户端对上述异常情况的拉取偏移量校对逻辑，必须正常校对拉取偏移量，否则消息消费将出现堆积。

Step5：如果待拉取偏移量大于 minOffset 并且小于 maxOffset，从当前 offset 处尝试拉取 32 条消息，根据消息队列偏移量（ConsumeQueue）从 commitlog 文件中查找消息在第 4 章中已经详细介绍了，在这里就不重复介绍了。

Step6：根据 PullResult 填充 responseHeader 的 nextBegionOffset、minOffset、maxOffset。

代码清单 5-19 PullMessageProcessor#processRequest

```
response.setRemark(getMessageResult.getStatus().name());
responseHeader.setNextBeginOffset(getMessageResult.getNextBeginOffset());
responseHeader.setMinOffset(getMessageResult.getMinOffset());
responseHeader.setMaxOffset(getMessageResult.getMaxOffset());
```

Step7：根据主从同步延迟，如果从节点数据包含下一次拉取的偏移量，设置下一次拉取任务的 brokerId。

Step8：根据 GetMessageResult 编码转换成关系，如表 5-1 所示。

表 5-1 GetMessageResult 与 Response 状态编码转换

	ResponseCode	GetMessageStatus
成功	SUCCESS	FOUND
立即重试	PULL_RETRY_IMMEDIATELY	MESSAGE_WAS_REMOVING：消息存放在下个 commitlog 文件中
偏移量移动	PULL_OFFSET_MOVED	NO_MATCHED_LOGIC_QUEUE：未找到队列 NO_MESSAGE_IN_QUEUE：队列中未包含消息 OFFSET_OVERFLOW_BADLY：offset 越界 OFFSET_TOO_SMALL：offset 未在消息队列中
未找到消息	PULL_NOT_FOUND	OFFSET_FOUND_NULL：消息物理偏移量为空 OFFSET_OVERFLOW_ONE：offset 越界一个

代码清单 5-20 PullMessageProcessor#processRequest

```
boolean storeOffsetEnable = brokerAllowSuspend;
storeOffsetEnable = storeOffsetEnable && hasCommitOffsetFlag;
storeOffsetEnable = storeOffsetEnable
        && this.brokerController.getMessageStoreConfig().getBrokerRole() !=
        BrokerRole.SLAVE;
if (storeOffsetEnable) {
    this.brokerController.getConsumerOffsetManager().commitOffset(RemotingHelp
        er.parseChannelRemoteAddr(channel),requestHeader.getConsumerGroup(),
        requestHeader.getTopic(), requestHeader.getQueueId(),
        requestHeader.getCommitOffset());
}
```

Step9：如果 commitlog 标记可用并且当前节点为主节点，则更新消息消费进度，消息消费进度详情在 5.6 节中重点讨论。

服务端消息拉取处理完毕，将返回结果到拉取消息调用方。在调用方，需要重点关注 PULL_RETRY_IMMEDIATELY、PULL_OFFSET_MOVED、PULL_NOT_FOUND 等情况下如何校正拉取偏移量。

3. 消息拉取客户端处理消息

回到消息拉取客户端调用入口：MQClientAPIImpl#pullMessageAsync，NettyRemotingClient 在收到服务端响应结构后会回调 PullCallback 的 onSuccess 或 onException，PullCallBack 对象在 DefaultMQPushConsumerImpl#pullMessage 中创建。

代码清单 5-21　MQClientAPIImpl#processPullResponse

```
PullStatus pullStatus = PullStatus.NO_NEW_MSG;
switch (response.getCode()) {
    case ResponseCode.SUCCESS:
        pullStatus = PullStatus.FOUND;
        break;
    case ResponseCode.PULL_NOT_FOUND:
        pullStatus = PullStatus.NO_NEW_MSG;
        break;
    case ResponseCode.PULL_RETRY_IMMEDIATELY:
        pullStatus = PullStatus.NO_MATCHED_MSG;
        break;
    case ResponseCode.PULL_OFFSET_MOVED:
        pullStatus = PullStatus.OFFSET_ILLEGAL;
        break;
    default:
        throw new MQBrokerException(response.getCode(), response.getRemark());
}
return new PullResultExt( 省略 PullResutExt 拉取结果的解码 );
```

Step1：根据响应结果解码成 PullResultExt 对象，此时只是从网络中读取消息列表到 byte[] messageBinary 属性。先重点看一下拉取状态码转换，如表 5-2 所示。

表 5-2　PullStatus 与 Response 状态编码转换

ResponseCode	PullStatus
SUCCESS	FOUND
PULL_RETRY_IMMEDIATELY	NO_MATCHED_MSG
PULL_OFFSET_MOVED	OFFSET_ILLEGAL
PULL_NOT_FOUND	NO_NEW_MSG

代码清单 5-22　DefaultMQPushConsumerImpl$PullCallBack#onSuccess

```
pullResult = DefaultMQPushConsumerImpl.this.pullAPIWrapper.processPullResult
    (pullRequest.getMessageQueue(), pullResult,subscriptionData);
```

Step2：调用 pullAPIWrapper 的 processPullResult 将消息字节数组解码成消息列表填充 msgFoundList，并对消息进行消息过滤（TAG）模式。PullResult 类图如图 5-8 所示。

```
┌─────────────────────────────────────────┐   ┌──────────────────────┐
│             PullResult                  │   │   <<enumeration>>    │
├─────────────────────────────────────────┤   │      PullStatus      │
│ -private final PullStatus pullStatus    │   ├──────────────────────┤
│ -private final long nextBeginOffset     │   │ FOUND                │
│ -private final long minOffset           │   │ NO_NEW_MSG           │
│ -private final long maxOffset           │   │ NO_MATCHED_MSG       │
│ -private List msgFoundList              │   │ OFFSET_ILLEGAL       │
└─────────────────────────────────────────┘   └──────────────────────┘
```

图 5-8　PullResult 类图

下面让我们一一来分析 PullResult 的核心属性。

1）pullStatus：拉取结果。
2）nextBeginOffset：下次拉取偏移量。
3）minOffset：消息队列最小偏移量。
4）maxOffset：消息队列最大偏移量。
5）msgFoundList：具体拉取的消息列表。

接下来先按照正常流程，即分析拉取结果为 PullStatus.FOUND（找到对应的消息）的情况来分析整个消息拉取过程。

代码清单 5-23　DefaultMQPushConsumerImpl$PullCallBack#onSuccess

```
long prevRequestOffset = pullRequest.getNextOffset();
pullRequest.setNextOffset(pullResult.getNextBeginOffset());
if (pullResult.getMsgFoundList() == null ||
        pullResult.getMsgFoundList().isEmpty()) {
    DefaultMQPushConsumerImpl.this.executePullRequestImmediately(pullRequest);}
}
```

Step3：更新 PullRequest 的下一次拉取偏移量，如果 msgFoundList 为空，则立即将 PullReqeuest 放入到 PullMessageService 的 pullRequestQueue，以便 PullMessageSerivce 能及时唤醒并再次执行消息拉取。为什么 PullStatus.FOUND,msgFoundList 还会为空呢？因为在 RocketMQ 根据 TAG 消息过滤，在服务端只是验证了 TAG 的 hashcode，在客户端再次对消息进行过滤，故可能会出现 msgFoundList 为空的情况。更多有关消息过滤的知识将在 5.8 节重点介绍。

代码清单 5-24　DefaultMQPushConsumerImpl$PullCallBack#onSuccess

```
boolean dispathToConsume =
            processQueue.putMessage(pullResult.getMsgFoundList());
DefaultMQPushConsumerImpl.this.consumeMessageService.
    submitConsumeRequest(pullResult.getMsgFoundList(),processQueue,
        pullRequest.getMessageQueue(),dispathToConsume);
```

Step4：首先将拉取到的消息存入 ProcessQueue，然后将拉取到的消息提交到 ConsumeMessageService 中供消费者消费，该方法是一个异步方法，也就是 PullCallBack 将消息提交到 ConsumeMessageService 中就会立即返回，至于这些消息如何消费，PullCallBack 不关注。

代码清单 5-25　DefaultMQPushConsumerImpl$PullCallBack#onSuccess

```
if (DefaultMQPushConsumerImpl.this.defaultMQPushConsumer.getPullInterval() > 0)
{
    DefaultMQPushConsumerImpl.this.executePullRequestLater(pullRequest,
        DefaultMQPushConsumerImpl.this.defaultMQPushConsumer
            .getPullInterval());
} else {
    DefaultMQPushConsumerImpl.this.executePullRequestImmediately(pullRequest);}
}
```

Step5：将消息提交给消费者线程之后 PullCallBack 将立即返回，可以说本次消息拉取顺利完成，然后根据 pullInterval 参数，如果 pullInterval>0，则等待 pullInterval 毫秒后将 PullRequest 对象放入到 PullMessageService 的 pullRequestQueue 中，该消息队列的下次拉取即将被激活，达到持续消息拉取，实现准实时拉取消息的效果。

再来分析消息拉取异常处理，如何校对拉取偏移量。

1）NO_NEW_MSG、NO_MATCHED_MSG

代码清单 5-26　DefaultMQPushConsumerImpl$PullCallBack#onSuccess

```
case NO_NEW_MSG:
case NO_MATCHED_MSG:
    pullRequest.setNextOffset(pullResult.getNextBeginOffset());
    DefaultMQPushConsumerImpl.this.correctTagsOffset(pullRequest);
    DefaultMQPushConsumerImpl.this.executePullRequestImmediately(pullRequest);
    break;
```

如果返回 NO_NEW_MSG（没有新消息）、NO_MATCHED_MSG（没有匹配消息），则直接使用服务器端校正的偏移量进行下一次消息的拉取。再来看看服务端是如何校正 Offset。

NO_NEW_MSG，对应 GetMessageResult.OFFSET_FOUND_NULL、GetMessageResult.OFFSET_OVERFLOW_ONE。

OFFSET_OVERFLOW_ONE：待拉取 offset 等于消息队列最大的偏移量，如果有新的消息到达，此时会创建一个新的 ConsumeQueue 文件，按照上一个 ConsueQueue 的最大偏移量就是下一个文件的起始偏移量，所以如果按照该 offset 第二次拉取消息时能成功。

OFFSET_FOUND_NULL：是根据 Consumequeue 的偏移量没有找到内容，将偏移量定位到下一个 ConsumeQueue，其实就是 offset +（一个 ConsumeQueue 包含多少个条目 =MappedFileSize / 20）。

2）OFFSET_ILLEGAL

代码清单 5-27　DefaultMQPushConsumerImpl$PullCallBack#onSuccess

```
pullRequest.setNextOffset(pullResult.getNextBeginOffset());
pullRequest.getProcessQueue().setDropped(true);
DefaultMQPushConsumerImpl.this.executeTaskLater(new Runnable() {
    public void run() {
        try {
            DefaultMQPushConsumerImpl.this.offsetStore.updateOffset (pullRequest.get
                MessageQueue(),pullRequest.getNextOffset(), false);
            DefaultMQPushConsumerImpl.this.offsetStore.persist
                (pullRequest.getMessageQueue());
            DefaultMQPushConsumerImpl.this.rebalanceImpl.removeProcessQueue
(pullReq
                uest.getMessageQueue());
        } catch (Throwable e) {
```

        }
      }
}, 10000);
```

如果拉取结果显示偏移量非法，首先将 ProcessQueue 设置 dropped 为 ture，表示丢弃该消费队列，意味着 ProcessQueue 中拉取的消息将停止消费，然后根据服务端下一次校对的偏移量尝试更新消息消费进度（内存中），然后尝试持久化消息消费进度，并将该消息队列从 RebalacnImpl 的处理队列中移除，意味着暂停该消息队列的消息拉取，等待下一次消息队列重新负载。OFFSET_ILLEGAL 对应服务端 GetMessageResult 状态的 NO_MATCHED_LOGIC_QUEUE、NO_MESSAGE_IN_QUEUE、OFFSET_OVERFLOW_BADLY、OFFSET_TOO_SMALL 中，这些状态服务端偏移量校正基本上使用原 offset，在客户端更新消息消费进度时只有当消息进度比当前消费进度大才会覆盖，保证消息进度的准确性。

RocketMQ 消息拉取比较复杂，其核心流程如图 5-9 所示。

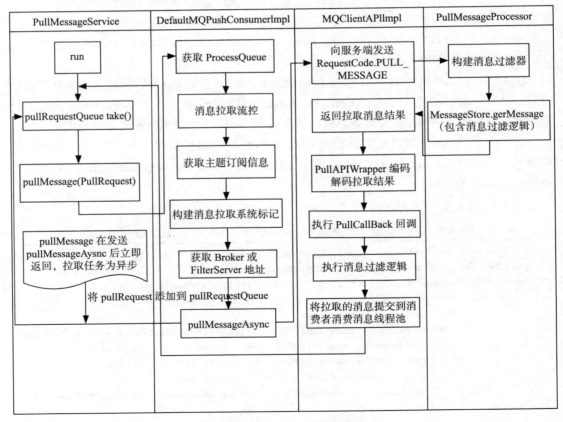

图 5-9 RocketMQ 消息拉取流程图
```

4. 消息拉取长轮询机制分析

RocketMQ 并没有真正实现推模式，而是消费者主动向消息服务器拉取消息，RocketMQ 推模式是循环向消息服务端发送消息拉取请求，如果消息消费者向 RocketMQ 发送消息拉取时，消息并未到达消费队列，如果不启用长轮询机制，则会在服务端等待 shortPolling-TimeMills 时间后（挂起）再去判断消息是否已到达消息队列，如果消息未到达则提示消息拉取客户端 PULL_NOT_FOUND（消息不存在），如果开启长轮询模式，RocketMQ 一方面会每 5s 轮询检查一次消息是否可达，同时一有新消息到达后立马通知挂起线程再次验证新消息是否是自己感兴趣的消息，如果是则从 commitlog 文件提取消息返回给消息拉取客户端，否则直到挂起超时，超时时间由消息拉取方在消息拉取时封装在请求参数中，PUSH 模式默认为 15s，PULL 模式通过 DefaultMQPullConsumer#setBrokerSuspendMaxTimeMillis 设置。RocketMQ 通过在 Broker 端配置 longPollingEnable 为 true 来开启长轮询模式。消息拉取时服务端从 Commitlog 未找到消息时的处理逻辑如下。

代码清单 5-28　PullMessageProcessor#processRequest

```
private RemotingCommand processRequest(final Channel channel,
            RemotingCommand request, boolean brokerAllowSuspend) {
    // 省略相关代码
    case ResponseCode.PULL_NOT_FOUND:
    if (brokerAllowSuspend && hasSuspendFlag) {
        long pollingTimeMills = suspendTimeoutMillisLong;
        if (!this.brokerController.getBrokerConfig().isLongPollingEnable()) {
            pollingTimeMills = this.brokerController.getBrokerConfig()
                .getShortPollingTimeMills();
        }
        String topic = requestHeader.getTopic();
        long offset = requestHeader.getQueueOffset();
        int queueId = requestHeader.getQueueId();
        PullRequest pullRequest = new PullRequest(request, channel,
            pollingTimeMills,this.brokerController.getMessageStore().now(),
            offset, subscriptionData, messageFilter);
        this.brokerController.getPullRequestHoldService().
            suspendPullRequest(topic, queueId, pullRequest);
        response = null;
        break;
    }
}
```

1）参数释义

Channel channel：网络通道，通过该通道向消息拉取客户端发送响应结果。

RemotingCommand request：消息拉取请求。

boolean brokerAllowSuspend：Broker 端是否支持挂起，处理消息拉取时默认传入 true，表示支持如果未找到消息则挂起，如果该参数为 false，未找到消息时直接返回客户端消息未找到。

2）如果 brokerAllowSuspend 为 true，表示支持挂起，则将响应对象 response 设置为 null，将不会立即向客户端写入响应，hasSuspendFlag 参数在拉取消息时构建的拉取标记，默认为 true。

3）默认支持挂起，则根据是否开启长轮询来决定挂起方式，如果支持长轮询模式，挂起超时时间来源于请求参数，PUSH 模式默认为 15s，PULL 模式通过 DefaultMQPullConsumer#brokerSuspenMaxTimeMillis 设置，默认 20s。然后创建拉取任务 PullRequest 并提交到 PullRequestHoldService 线程中。

RocketMQ 轮询机制由两个线程共同来完成。

1）PullRequestHoldService：每隔 5s 重试一次。

2）DefaultMessageStore#ReputMessageService，每处理一次重新拉取，Thread.sleep（1），继续下一次检查。

5. PullRequestHoldService 线程详解

代码清单 5-29　PullRequestHoldService#suspendPullRequest

```
public void suspendPullRequest(final String topic, final int queueId, final
        PullRequest pullRequest)
    String key = this.buildKey(topic, queueId);
    ManyPullRequest mpr = this.pullRequestTable.get(key);
    if (null == mpr) {
        mpr = new ManyPullRequest();
        ManyPullRequest prev = this.pullRequestTable.putIfAbsent(key, mpr);
        if (prev != null) {
            mpr = prev;
        }
    }
    mpr.addPullRequest(pullRequest);
}
```

根据消息主题与消息队列构建 key，从 ConcurrentMap<String/* topic@queueId */, ManyPullRequest> pullRequestTable 中获取该主题 @ 队列对应的 ManyPullRequest，通过 ConcurrentMap 的并发特性，维护主题 @ 队列的 ManyPullRequest，然后将 PullRequest 放入 ManyPullRequest。ManyPullRequest 对象内部持有一个 PullRequest 列表，表示同一主题 @ 队列的累积拉取消息任务。

代码清单 5-30　PullRequestHoldService#run

```
public void run() {
    log.info("{} service started", this.getServiceName());
    while (!this.isStopped()) {
        try {
            if (this.brokerController.getBrokerConfig().isLongPollingEnable()) {
                this.waitForRunning(5 * 1000);
            } else {
```

```
                this.waitForRunning(this.brokerController.getBrokerConfig().
                    getShortPollingTimeMills());
            }
            long beginLockTimestamp = this.systemClock.now();
            this.checkHoldRequest();
            long costTime = this.systemClock.now() - beginLockTimestamp;
            if (costTime > 5 * 1000) {
                log.info("[NOTIFYME] check hold request cost {} ms.", costTime);
            }
        } catch (Throwable e) {
            log.warn(this.getServiceName() + " service has exception. ", e);
        }
    }
    log.info("{} service end", this.getServiceName());
}
```

如果开启长轮询，每 5s 尝试一次，判断新消息是否到达。如果未开启长轮询，则默认等待 1s 再次尝试，可以通过 BrokerConfig#shortPollingTimeMills 改变等待时间。PullRequestHold Service 的核心逻辑见 PullRequestHoldService#checkHoldRequest。

代码清单 5-31　PullRequestHoldService#checkHoldRequest

```
private void checkHoldRequest() {
    for (String key : this.pullRequestTable.keySet()) {
        String[] kArray = key.split(TOPIC_QUEUEID_SEPARATOR);
        if (2 == kArray.length) {
            String topic = kArray[0];
            int queueId = Integer.parseInt(kArray[1]);
            final long offset = this.brokerController.getMessageStore().
                getMaxOffsetInQueue(topic, queueId);
            try {
                this.notifyMessageArriving(topic, queueId, offset);
            } catch (Throwable e) {
                log.error("check hold request failed. topic={}, queueId={}", topic,
                    queueId, e);
            }
        }
    }
}
```

遍历拉取任务表，根据主题与队列获取消息消费队列最大偏移量，如果该偏移量大于待拉取偏移量，说明有新的消息到达，调用 notifyMessageArriving 触发消息拉取。

PullRequestHoldService#notifyMessageArriving 详解如下。

代码清单 5-32　PullRequestHoldService#notifyMessageArriving

```
List<PullRequest> requestList = mpr.cloneListAndClear();
public synchronized List<PullRequest> cloneListAndClear() {
    if (!this.pullRequestList.isEmpty()) {
        List<PullRequest> result = (ArrayList<PullRequest>)
```

```
            this.pullRequestList.clone();
        this.pullRequestList.clear();
        return result;
    }
    return null;
}
```

Step1：首先从 ManyPullRequest 中获取当前该主题、队列所有的挂起拉取任务。值得注意的是该方法使用了 synchronized，说明该数据结构会存在并发访问，该属性是 PullRequestHoldService 线程的私有属性，会存在并发？答案是存在并发，下文重点提到的 ReputMessageService 内部将持有 PullRequestHoldService，也会唤醒挂起线程从而执行消息拉取尝试。

代码清单 5-33　PullRequestHoldService#notifyMessageArriving

```
if (newestOffset > request.getPullFromThisOffset()) {
    boolean match = request.getMessageFilter().isMatchedByConsumeQueue(tagsCode,
        new ConsumeQueueExt.CqExtUnit(tagsCode, msgStoreTime, filterBitMap));
    if (match && properties != null) {
        match = request.getMessageFilter().isMatchedByCommitLog(null, properties);
    }
    if (match) {
        try {
            this.brokerController.getPullMessageProcessor().
                executeRequestWhenWakeup(request.getClientChannel(),request.getRe
                    questCommand());
        } catch (Throwable e) {
            log.error("execute request when wakeup failed.", e);
        }
        continue;
    }
}
```

Step2：如果消息队列的最大偏移量大于待拉取偏移量，如果消息匹配则调用 executeRequest WhenWakeup 将消息返回给消息拉取客户端，否则等待下一次尝试。

代码清单 5-34　PullRequestHoldService#notifyMessageArriving

```
if (System.currentTimeMillis() >= (request.getSuspendTimestamp() +
        request.getTimeoutMillis())) {
    try {
        this.brokerController.getPullMessageProcessor().
            executeRequestWhenWakeup(request.getClientChannel(),
                request.getRequestCommand());
    } catch (Throwable e) {
        log.error("execute request when wakeup failed.", e);
    }
    continue;
}
```

Step3：如果挂起超时时间超时，则不继续等待将直接返回客户消息未找到。

代码清单 5-35　PullMessageProcessor#executeRequestWhenWakeup

```
final RemotingCommand response = PullMessageProcessor.this.
    processRequest(channel, request, false);
```

Step4：这里的核心又回到长轮询的入口代码了，其核心是设置 brokerAllowSuspend 为 false，表示不支持拉取线程挂起，即当根据偏移量无法获取消息时将不挂起线程等待新消息到来，而是直接返回告诉客户端本次消息拉取未找到消息。

回想一下，如果当开启了长轮询机制，PullRequestHoldService 线程会每隔 5s 被唤醒去尝试检测是否有新消息的到来直到超时，如果被挂起，需要等待 5s，消息拉取实时性比较差，为了避免这种情况，RocketMQ 引入另外一种机制：当消息到达时唤醒挂起线程触发一次检查。

6. DefaultMessageStore$ReputMessageService 详解

ReputMessageService 线程主要是根据 Commitlog 将消息转发到 ConsumeQueue、Index 等文件，该部分已经在第 4 章做了详细解读，本节关注 doReput 方法关于长轮询相关实现。

代码清单 5-36　DefaultMessageStore#start

```
if (this.getMessageStoreConfig().isDuplicationEnable()) {
    this.reputMessageService.setReputFromOffset(this.commitLog.getConfirmOffset());
} else {
    this.reputMessageService.setReputFromOffset(this.commitLog.getMaxOffset());
}
this.reputMessageService.start();
```

如果允许消息重复，设置重新推送偏移量为 Commitlog 文件的提交偏移量，如果不允许重复推送则设置重新推送偏移为 commitlog 当前最大的偏移量。

代码清单 5-37　DefaultMessageStore$ReputMessageService#doReput

```
if (BrokerRole.SLAVE !=
        DefaultMessageStore.this.getMessageStoreConfig().getBrokerRole() &&
        DefaultMessageStore.this.brokerConfig.isLongPollingEnable()){
    DefaultMessageStore.this.messageArrivingListener.arriving(
        dispatchRequest.getTopic(),
        dispatchRequest.getQueueId(),
        dispatchRequest.getConsumeQueueOffset() + 1,
        dispatchRequest.getTagsCode(), dispatchRequest.getStoreTimestamp(),
        dispatchRequest.getBitMap(), dispatchRequest.getPropertiesMap());
}
```

当新消息达到 CommitLog 时，ReputMessageService 线程负责将消息转发给 ConsumeQueue、IndexFile，如果 Broker 端开启了长轮询模式并且角色主节点，则最终将调用 PullRequestHoldService 线程的 notifyMessageArriving 方法唤醒挂起线程，判断当前消费队列最大偏移量是否大于待拉取偏移量，如果大于则拉取消息。长轮询模式使得消息拉取能实现准实时。

5.5 消息队列负载与重新分布机制

PullMessageService 在启动时由于 LinkedBlockingQueue<PullRequest> pullRequestQueue 中没有 PullRequest 对象，故 PullMessageService 线程将阻塞。

问题 1：PullRequest 对象在什么时候创建并加入到 pullRequestQueue 中以便唤醒 PullMessageService 线程。

问题 2：集群内多个消费者是如何负载主题下的多个消费队列，并且如果有新的消费者加入时，消息队列又会如何重新分布。

RocketMQ 消息队列重新分布是由 RebalanceService 线程来实现的。一个 MQClientInstance 持有一个 RebalanceService 实现，并随着 MQClientInstance 的启动而启动。

带着上面两个问题走入 RebalanceService 的 run 方法中。

代码清单 5-38　RebalanceService#run

```
public void run() {
    log.info(this.getServiceName() + " service started");
    while (!this.isStopped()) {
        this.waitForRunning(waitInterval);
        this.mqClientFactory.doRebalance();
    }
    log.info(this.getServiceName() + " service end");
}
```

RebalanceService 线程默认每隔 20s 执行一次 mqClientFactory.doRebalance() 方法，可以使用 -Drocketmq.client.rebalance.waitInterval=interval 来改变默认值。

代码清单 5-39　MQClientInstance#doRebalance

```
public void doRebalance() {
    for (Map.Entry<String, MQConsumerInner> entry :
        this.consumerTable.entrySet()) {
        MQConsumerInner impl = entry.getValue();
        if (impl != null) {
            try {
                impl.doRebalance();
            } catch (Throwable e) {
                log.error("doRebalance exception", e);
            }
        }
    }
}
```

MQClientIinstance 遍历已注册的消费者，对消费者执行 doRebalance() 方法。

代码清单 5-40　RebalanceImpl#doRebalance

```
public void doRebalance(final boolean isOrder) {
    Map<String, SubscriptionData> subTable = this.getSubscriptionInner();
    if (subTable != null) {
        for (final Map.Entry<String, SubscriptionData> entry : subTable.entrySet()) {
            final String topic = entry.getKey();
            try {
                this.rebalanceByTopic(topic, isOrder);
            } catch (Throwable e) {
                if (!topic.startsWith(MixAll.RETRY_GROUP_TOPIC_PREFIX)) {
                    log.warn("rebalanceByTopic Exception", e);
                }
            }
        }
    }
    this.truncateMessageQueueNotMyTopic();
}
```

每个 DefaultMQPushConsumerImpl 都持有一个单独的 RebalanceImpl 对象，该方法主要是遍历订阅信息对每个主题的队列进行重新负载。RebalanceImpl 的 Map<String, SubscriptionData> subTable 在调用消费者 DefaultMQPushConsumerImpl#subscribe 方法时填充。如果订阅信息发送变化，例如调用了 unsubscribe 方法，则需要将不关心的主题消费队列从 processQueueTable 中移除。接下来重点分析 RebalanceImpl#rebalanceByTopic 来分析 RocketMQ 是如何针对单个主题进行消息队列重新负载（以集群模式）。

代码清单 5-41　RebalanceImpl#rebalanceByTopic

```
Set<MessageQueue> mqSet = this.topicSubscribeInfoTable.get(topic);
List<String> cidAll = this.mQClientFactory.findConsumerIdList(topic,
    consumerGroup);
```

Step1：从主题订阅信息缓存表中获取主题的队列信息；发送请求从 Broker 中该消费组内当前所有的消费者客户端 ID，主题 topic 的队列可能分布在多个 Broker 上，那请求发往哪个 Broker 呢？RocketMQ 从主题的路由信息表中随机选择一个 Broker。Broker 为什么会存在消费组内所有消费者的信息呢？我们不妨回忆一下消费者在启动的时候会向 MQClientInstance 中注册消费者，然后 MQClientInstance 会向所有的 Broker 发送心跳包，心跳包中包含 MQClientInstance 的消费者信息。如果 mqSet、cidAll 任意一个为空则忽略本次消息队列负载。

代码清单 5-42　RebalanceImpl#rebalanceByTopic

```
Collections.sort(mqAll);
Collections.sort(cidAll);
AllocateMessageQueueStrategy strategy = this.allocateMessageQueueStrategy;
```

```
List<MessageQueue> allocateResult = null;
    try {
        allocateResult = strategy.allocate(
            this.consumerGroup,this.mQClientFactory.getClientId(),mqAll,cidAll);
    } catch (Throwable e) {
        log.error("AllocateMessageQueueStrategy.allocate Exception. 
            allocateMessageQueueStrategyName={}",strategy.getName(),e);
        return;
    }
```

Step2：首先对 cidAll,mqAll 排序，这个很重要，同一个消费组内看到的视图保持一致，确保同一个消费队列不会被多个消费者分配。RocketMQ 消息队列分配算法接口。

代码清单 5-43　AllocateMessageQueueStrategy

```
/**
 * Strategy Algorithm for message allocating between consumers
 */
public interface AllocateMessageQueueStrategy {
    List<MessageQueue> allocate(String consumerGroup,
        String currentCID,List<MessageQueue> mqAll,List<String> cidAll
    );
    /**
     * Algorithm name
     *
     * @return The strategy name
     */
    String getName();
}
```

RocketMQ 默认提供 5 种分配算法。

1）AllocateMessageQueueAveragely：平均分配，推荐指数为 5 颗星。

举例来说，如果现在有 8 个消息消费队列 q1,q2,q3,q4,q5,q6,q7,q8，有 3 个消费者 c1,c2,c3，那么根据该负载算法，消息队列分配如下：

c1: q1,q2,q3

c2:q4,q5,q6

c3:q7,q8

2）AllocateMessageQueueAveragelyByCircle：平均轮询分配，推荐指数为 5 颗星。

举例来说，如果现在有 8 个消息消费队列 q1,q2,q3,q4,q5,q6,q7,q8，有 3 个消费者 c1,c2,c3，那么根据该负载算法，消息队列分配如下：

c1：q1,q4,q7

c2：q2,q5,q8

c3：q3,q6

3）AllocateMessageQueueConsistentHash：一致性 hash。不推荐使用，因为消息队列负

载信息不容易跟踪。

4）AllocateMessageQueueByConfig：根据配置，为每一个消费者配置固定的消息队列。

5）AllocateMessageQueueByMachineRoom：根据 Broker 部署机房名，对每个消费者负责不同的 Broker 上的队列。

 注意　消息负载算法如果没有特殊的要求，尽量使用 AllocateMessageQueueAveragely、AllocateMessageQueueAveragelyByCircle，因为分配算法比较直观。消息队列分配遵循一个消费者可以分配多个消息队列，但同一个消息队列只会分配给一个消费者，故如果消费者个数大于消息队列数量，则有些消费者无法消费消息。

对比消息队列是否发生变化，主要思路是遍历当前负载队列集合，如果队列不在新分配队列集合中，需要将该队列停止消费并保存消费进度；遍历已分配的队列，如果队列不在队列负载表中（processQueueTable）则需要创建该队列拉取任务 PullRequest，然后添加到 PullMessageService 线程的 pullRequestQueue 中，PullMessageService 才会继续拉取任务。

代码清单 5-44　RebalanceImpl#updateProcessQueueTableInRebalance

```
Iterator<Entry<MessageQueue, ProcessQueue>> it =
                    this.processQueueTable.entrySet().iterator();
while (it.hasNext()) {
    Entry<MessageQueue, ProcessQueue> next = it.next();
    MessageQueue mq = next.getKey();
    ProcessQueue pq = next.getValue();
    if (mq.getTopic().equals(topic)) {
        if (!mqSet.contains(mq)) {
            pq.setDropped(true);
            if (this.removeUnnecessaryMessageQueue(mq, pq)) {
                it.remove();
                changed = true;
                log.info("doRebalance, {}, remove unnecessary mq, {}",
                    consumerGroup, mq);
            }
        }
    }
}
```

Step3：ConcurrentMap<MessageQueue, ProcessQueue> processQueueTable，当前消费者负载的消息队列缓存表，如果缓存表中的 MessageQueue 不包含在 mqSet 中，说明经过本次消息队列负载后，该 mq 被分配给其他消费者，故需要暂停该消息队列消息的消费，方法是将 ProccessQueue 的状态设置为 droped=true，该 ProcessQueue 中的消息将不会再被消费，调用 removeUnnecessaryMessageQueue 方法判断是否将 MessageQueue、ProccessQueue 缓存表中移除。removeUnnecessaryMessageQueue 在 RebalanceImple 定义为抽象方法。removeUnnecessaryMessageQueue 方法主要持久化待移除 MessageQueue 消息消费进度。在 Push

模式下，如果是集群模式并且是顺序消息消费时，还需要先解锁队列，关于顺序消息将在5.9节详细讨论。

代码清单5-45　RebalanceImpl#updateProcessQueueTableInRebalance

```
List<PullRequest> pullRequestList = new ArrayList<PullRequest>();
for (MessageQueue mq : mqSet) {
    if (!this.processQueueTable.containsKey(mq)) {
        if (isOrder && !this.lock(mq)) {
            log.warn("doRebalance, {}, add a new mq failed, {}, because lock
                failed", consumerGroup, mq);
            continue;
        }
        this.removeDirtyOffset(mq);
        ProcessQueue pq = new ProcessQueue();
        long nextOffset = this.computePullFromWhere(mq);
        if (nextOffset >= 0) {
            ProcessQueue pre = this.processQueueTable.putIfAbsent(mq, pq);
            if (pre != null) {
                log.info("doRebalance, {}, mq already exists, {}", consumerGroup,
                    mq);
            } else {
                log.info("doRebalance, {}, add a new mq, {}", consumerGroup, mq);
                PullRequest pullRequest = new PullRequest();
                pullRequest.setConsumerGroup(consumerGroup);
                pullRequest.setNextOffset(nextOffset);
                pullRequest.setMessageQueue(mq);
                pullRequest.setProcessQueue(pq);
                pullRequestList.add(pullRequest);
                changed = true;
            }
        } else {
            log.warn("doRebalance, {}, add new mq failed, {}", consumerGroup,
                mq);
        }
    }
}
```

Step4：遍历本次负载分配到的队列集合，如果processQueueTable中没有包含该消息队列，表明这是本次新增加的消息队列，首先从内存中移除该消息队列的消费进度，然后从磁盘中读取该消息队列的消费进度，创建PullRequest对象。这里有一个关键，如果读取到的消费进度小于0，则需要校对消费进度。RocketMQ提供CONSUME_FROM_LAST_OFFSET、CONSUME_FROM_FIRST_OFFSET、CONSUME_FROM_TIMESTAMP方式，在创建消费者时可以通过调用DefaultMQPushConsumer#setConsumeFromWhere方法设置。PullRequest的nextOffset计算逻辑位于：RebalancePushImpl#computePullFromWhere。

1）ConsumeFromWhere.CONSUME_FROM_LAST_OFFSET：从队列最新偏移量开始消费。

代码清单 5-46　RebalancePushImpl#computePullFromWhere

```
case CONSUME_FROM_LAST_OFFSET: {
    long lastOffset = offsetStore.readOffset(mq, ReadOffsetType.READ_FROM_STORE);
    if (lastOffset >= 0) {
        result = lastOffset;
    } else if (-1 == lastOffset) { // First start,no offset
        if (mq.getTopic().startsWith(MixAll.RETRY_GROUP_TOPIC_PREFIX)) {
            result = 0L;
        } else {
            try {
                result = this.mQClientFactory.getMQAdminImpl().maxOffset(mq);
            } catch (MQClientException e) {
                result = -1;
            }
        }
    } else {
        result = -1;
    }
    break;
}
```

offsetStore.readOffset（mq, ReadOffsetType.READ_FROM_STORE）返回 -1 表示该消息队列刚创建。从磁盘中读取到消息队列的消费进度，如果大于 0 则直接返回即可；如果等于 -1，CONSUME_FROM_LAST_OFFSET 模式下获取该消息队列当前最大的偏移量。如果小于 -1，则表示该消息进度文件中存储了错误的偏移量，返回 -1。

2）CONSUME_FROM_FIRST_OFFSET：从头开始消费。

代码清单 5-47　RebalancePushImpl#computePullFromWhere

```
case CONSUME_FROM_FIRST_OFFSET: {
    long lastOffset = offsetStore.readOffset(mq, ReadOffsetType.READ_FROM_STORE);
    if (lastOffset >= 0) {
        result = lastOffset;
    } else if (-1 == lastOffset) {
        result = 0L;
    } else {
        result = -1;
    }
    break;
}
```

从磁盘中读取到消息队列的消费进度，如果大于 0 则直接返回即可；如果等于 -1，CONSUME_FROM_FIRST_OFFSET 模式下直接返回 0，从头开始。如果小于 -1，则表示该消息进度文件中存储了错误的偏移量，返回 -1。

3）CONSUME_FROM_TIMESTAMP：从消费者启动的时间戳对应的消费进度开始消费。

```
        try {
            long lastOffset = offsetStore.readOffset(mq, ReadOffsetType.READ_FROM_STORE);
            if (lastOffset >= 0) {
                result = lastOffset;
            } else if (-1 == lastOffset) {
                try {
                    long timestamp = UtilAll.parseDate(this.defaultMQPushConsumerImpl
                            .getDefaultMQPushConsumer().getConsumeTimestamp(),
                            UtilAll.YYYYMMDDHHMMSS).getTime();
                    result = this.mQClientFactory.getMQAdminImpl().searchOffset(mq,
                            timestamp);
                } catch (MQClientException e) {
                    result = -1;
                }
            } else {
                result = -1;
            }
```

从磁盘中读取到消息队列的消费进度，如果大于 0 则直接返回即可；如果等于 -1，CONSUME_FROM_TIMESTAMP 模式下会尝试去操作消息存储时间戳为消费者启动的时间戳，如果能找到则返回找到的偏移量，否则返回 0。如果小于 -1，则表示该消息进度文件中存储了错误的偏移量，返回 -1。

> **注意** ConsumeFromWhere 相关消费进度校正策略只有在从磁盘中获取消费进度返回 -1 时才会生效，如果从消息进度存储文件中返回的消费进度小于 -1，表示偏移量非法，则使用偏移量 -1 去拉取消息，那么会发生什么呢？首先第一次去消息服务器拉取消息时无法取到消息，但是会用 -1 去更新消费进度，然后将消息消费队列丢弃，在下一次消息队列负载时会再次消费。

代码清单 5-48　RebalancePushImpl#dispatchPullRequest

```
this.dispatchPullRequest(pullRequestList);
public void dispatchPullRequest(List<PullRequest> pullRequestList) {
    for (PullRequest pullRequest : pullRequestList) {
        this.defaultMQPushConsumerImpl.executePullRequestImmediately(pullRequest);
        log.info("doRebalance, {}, add a new pull request {}", consumerGroup,
            pullRequest);
    }
}
```

Step5：将 PullRequest 加入到 PullMessageService 中，以便唤醒 PullMessageService 线程。

消息队列负载机制就介绍到这，回到本节的两个问题。

问题 1：PullRequest 对象在什么时候创建并加入到 pullRequestQueue 中以便唤醒 PullMessageService 线程。

RebalanceService 线程每隔 20s 对消费者订阅的主题进行一次队列重新分配，每一次分配都会获取主题的所有队列、从 Broker 服务器实时查询当前该主题该消费组内消费者列表，对新分配的消息队列会创建对应的 PullRequest 对象。在一个 JVM 进程中，同一个消费组同一个队列只会存在一个 PullRequest 对象。

问题 2：集群内多个消费者是如何负载主题下的多个消费队列，并且如果有新的消费者加入时，消息队列又会如何重新分布。

由于每次进行队列重新负载时会从 Broker 实时查询出当前消费组内所有消费者，并且对消息队列、消费者列表进行排序，这样新加入的消费者就会在队列重新分布时分配到消费队列从而消费消息。

本节分析了消息队列的负载机制，结合 5.4 节，RocketMQ 消息拉取由 PullMessage-Service 与 RebalanceService 共同协作完成，如图 5-10 所示。

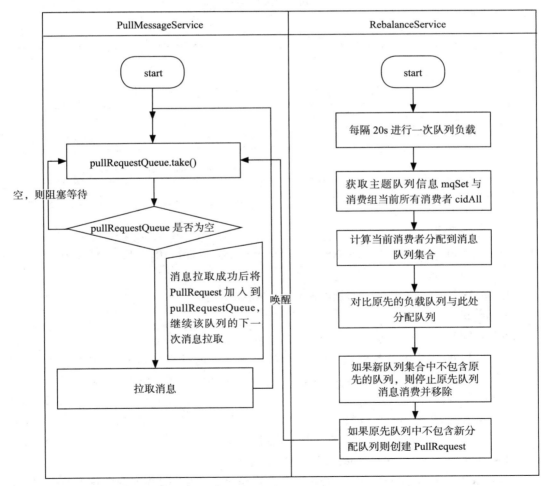

图 5-10　PullMessageService 线程与 RebalanceService 线程交互图

5.6 消息消费过程

回顾一下消息拉取，PullMessageService 负责对消息队列进行消息拉取，从远端服务器拉取消息后将消息存入 ProcessQueue 消息队列处理队列中，然后调用 ConsumeMessageService#submitConsumeRequest 方法进行消息消费，使用线程池来消费消息，确保了消息拉取与消息消费的解耦。RocketMQ 使用 ConsumeMessageService 来实现消息消费的处理逻辑。RocketMQ 支持顺序消费与并发消费，本节将重点关注并发消费的消费流程，顺序消费将在 5.9 节中详细分析。ConsumeMessageService 核心类图如图 5-11 所示。

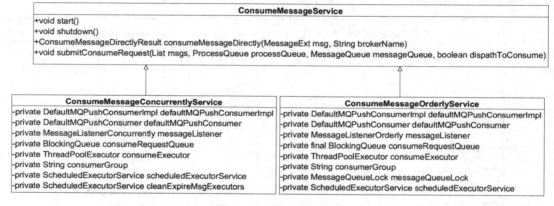

图 5-11 消息消费类图

核心方法描述如下。

1）ConsumeMessageDirectlyResult consumeMessageDirectly（final MessageExt msg,final String brokerName）

直接消费消息，主要用于通过管理命令收到消费消息。

MessageExt ms：消息

borkerName：Broker 名称。

2）void submitConsumeRequest（final List\<MessageExt\> msgs,final ProcessQueue processQueue,final MessageQueue messageQueue,final boolean dispathToConsume）

提交消息消费

List\<MessageExt\> msgs：消息列表，默认一次从服务器最多拉取 32 条。

ProcessQueue processQueue：消息处理队列。

MessageQueue messageQueue：消息所属消费队列。

boolean dispathToConsume：是否转发到消费线程池，并发消费时忽略该参数。

ConsumeMessageConcurrentlyService（并发消息消费核心参数一览）：

1）DefaultMQPushConsumerImpl defaultMQPushConsumerImpl：消息推模式实现类。
2）DefaultMQPushConsumer defaultMQPushConsumer：消费者对象。
3）MessageListenerConcurrently messageListener：并发消息业务事件类。
4）BlockingQueue<Runnable> consumeRequestQueue：消息消费任务队列。
5）ThreadPoolExecutor consumeExecutor：消息消费线程池。
6）String consumerGroup：消费组。
7）ScheduledExecutorService scheduledExecutorService：添加消费任务到consumeExecutor延迟调度器。
8）ScheduledExecutorService cleanExpireMsgExecutors：定时删除过期消息线程池。

为了揭示消息消费的完整过程，从服务器拉取到消息后回调 PullCallBack 回调方法后，先将消息放入到 ProccessQueue 中，然后把消息提交到消费线程池中执行，也就是调用 ConsumeMessageService#submitConsumeRequest 开始进入到消息消费的世界中来。

5.6.1 消息消费

消费者消息消费服务 ConsumeMessageConcurrentlyService 的主要方法是 submitConsumeRequest 提交消费请求，具体逻辑如下。

代码清单 5-49　ConsumeMessageConcurrentlyService#submitConsumeRequest

```
final int consumeBatchSize =
      this.defaultMQPushConsumer.getConsumeMessageBatchMaxSize();
if (msgs.size() <= consumeBatchSize) {
    ConsumeRequest consumeRequest = new ConsumeRequest(msgs, processQueue,
        messageQueue);
    try {
        this.consumeExecutor.submit(consumeRequest);
    } catch (RejectedExecutionException e) {
        this.submitConsumeRequestLater(consumeRequest);
    }
}
```

Step1：consumeMessageBatchMaxSize，消息批次，在这里看来也就是一次消息消费任务 ConsumeRequest 中包含的消息条数，默认为1，msgs.size() 默认最多为32条，受 DefaultMQPushConsumer.pullBatchSize 属性控制，如果 msgs.size() 小于 consumeMessageBatchMaxSize，则直接将拉取到的消息放入到 ConsumeRequest 中，然后将 consumeRequest 提交到消息消费者线程池中，如果提交过程中出现拒绝提交异常则延迟 5s 再提交，这里其实是给出一种标准的拒绝提交实现方式，实际过程中由于消费者线程池使用的任务队列为 LinkedBlockingQueue 无界队列，故不会出现拒绝提交异常。

代码清单 5-50　ConsumeMessageConcurrentlyService#submitConsumeRequest

```
if (msgs.size() > consumeBatchSize) {
```

```
        for (int total = 0; total < msgs.size(); ) {
            List<MessageExt> msgThis = new ArrayList<MessageExt>(consumeBatchSize);
            for (int i = 0; i < consumeBatchSize; i++, total++) {
                if (total < msgs.size()) {
                    msgThis.add(msgs.get(total));
                } else {
                    break;
                }
            }
            ConsumeRequest consumeRequest = new ConsumeRequest(msgThis,
                processQueue,messageQueue);
            try {
                this.consumeExecutor.submit(consumeRequest);
            } catch (RejectedExecutionException e) {
                for (; total < msgs.size(); total++) {
                    msgThis.add(msgs.get(total));
                }
                this.submitConsumeRequestLater(consumeRequest);
            }
        }
    }
```

Step2：如果拉取的消息条数大于 consumeMessageBatchMaxSize，则对拉取消息进行分页，每页 consumeMessageBatchMaxSize 条消息，创建多个 ConsumeRequest 任务并提交到消费线程池。ConsumeRequest 的 run 方法封装了具体消息消费逻辑。

代码清单 5-51　ConsumeMessageConcurrentlyService$ConsumeRequest#run

```
if (this.processQueue.isDropped()) {
    log.info("the message queue not be able to consume, because it's dropped.
        group={} {}", ConsumeMessageConcurrentlyService.this.consumerGroup,
            this.messageQueue);
    return;
}
```

Step3：进入具体消息消费时会先检查 processQueue 的 dropped，如果设置为 true，则停止该队列的消费，在进行消息重新负载时如果该消息队列被分配给消费组内其他消费者后，需要 droped 设置为 true，阻止消费者继续消费不属于自己的消息队列。

Step4：执行消息消费钩子函数 ConsumeMessageHook#consumeMessageBefore 函数，通过 consumer.getDefaultMQPushConsumerImpl().registerConsumeMessageHook（hook）方法消息消费执行钩子函数。

代码清单 5-52　ConsumeMessageConcurrentlyService#resetRetryTopic

```
public void resetRetryTopic(final List<MessageExt> msgs) {
    final String groupTopic = MixAll.getRetryTopic(consumerGroup);
    for (MessageExt msg : msgs) {
        String retryTopic = msg.getProperty(MessageConst.PROPERTY_RETRY_TOPIC);
```

```
        if (retryTopic != null && groupTopic.equals(msg.getTopic())) {
            msg.setTopic(retryTopic);
        }
    }
}
```

Step5：恢复重试消息主题名。这是为什么呢？这是由消息重试机制决定的，RocketMQ 将消息存入 commitlog 文件时，如果发现消息的延时级别 delayTimeLevel 大于 0，会首先将重试主题存入在消息的属性中，然后设置主题名称为 SCHEDULE_TOPIC，以便时间到后重新参与消息消费。

代码清单 5-53　ConsumeMessageConcurrentlyService$ConsumeRequest#run

```
try {
    ConsumeMessageConcurrentlyService.this.resetRetryTopic(msgs);
    if (msgs != null && !msgs.isEmpty()) {
        for (MessageExt msg : msgs) {
            MessageAccessor.setConsumeStartTimeStamp(msg,
                String.valueOf(System.currentTimeMillis()));
        }
    }
    status = listener.consumeMessage(Collections.unmodifiableList(msgs),
        context);
} catch (Throwable e) {
    hasException = true;
}
```

Step6：执行具体的消息消费，调用应用程序消息监听器的 consumeMessage 方法，进入到具体的消息消费业务逻辑，返回该批消息的消费结果。最终将返回 CONSUME_SUCCESS（消费成功）或 RECONSUME_LATER（需要重新消费）。

Step7：执行消息消费钩子函数 ConsumeMessageHook#consumeMessageAfter 函数。

代码清单 5-54　ConsumeMessageConcurrentlyService$ConsumeRequest#run

```
if (!processQueue.isDropped()) {
    ConsumeMessageConcurrentlyService.this.processConsumeResult(status,
        context, this);
}
```

Step8：执行业务消息消费后，在处理结果前再次验证一下 ProcessQueue 的 isDroped 状态值，如果设置为 true，将不对结果进行处理，也就是说如果在消息消费过程中进入到 Step4 时，如果由于由新的消费者加入或原先的消费者出现宕机导致原先分给消费者的队列在负载之后分配给别的消费者，那么在应用程序的角度来看的话，消息会被重复消费。

代码清单 5-55　ConsumeMessageConcurrentlyService#processConsumeResult

```
switch (status) {
```

```java
        case CONSUME_SUCCESS:
            if (ackIndex >= consumeRequest.getMsgs().size()) {
                ackIndex = consumeRequest.getMsgs().size() - 1;
            }
            break;
        case RECONSUME_LATER:
            ackIndex = -1;
            break;
        default:
            break;
    }
```

Step9：根据消息监听器返回的结果，计算 ackIndex，如果返回 CONSUME_SUCCESS，ackIndex 设置为 msgs.size()-1，如果返回 RECONSUME_LATER，ackIndex=-1，这是为下文发送 msg back（ACK）消息做准备的。

代码清单 5-56　ConsumeMessageConcurrentlyService#processConsumeResult

```java
switch (this.defaultMQPushConsumer.getMessageModel()) {
    case BROADCASTING:
        for (int i = ackIndex + 1; i < consumeRequest.getMsgs().size(); i++) {
            MessageExt msg = consumeRequest.getMsgs().get(i);
            log.warn("BROADCASTING, the message consume failed, drop it, {}",
                msg.toString());
        }
        break;
    case CLUSTERING:
        List<MessageExt> msgBackFailed = new ArrayList<MessageExt>
                (consumeRequest.getMsgs().size());
        for (int i = ackIndex + 1; i < consumeRequest.getMsgs().size(); i++) {
            MessageExt msg = consumeRequest.getMsgs().get(i);
            boolean result = this.sendMessageBack(msg, context);
            if (!result) {
                msg.setReconsumeTimes(msg.getReconsumeTimes() + 1);
                msgBackFailed.add(msg);
            }
        }

        if (!msgBackFailed.isEmpty()) {
            consumeRequest.getMsgs().removeAll(msgBackFailed);
            this.submitConsumeRequestLater(msgBackFailed,
                consumeRequest.getProcessQueue(), consumeRequest.getMessageQueue());
        }
        break;
    default:
        break;
}
```

Step10：如果是集群模式，业务方返回 RECONSUME_LATER，消息并不会重新被消费，只是以警告级别输出到日志文件。如果是集群模式，消息消费成功，由于 ackIndex= consumeRequest.getMsgs().size()-1，故 i=ackIndex+1 等 于 consumeRequest.getMsgs().size()，并不会执行 sendMessageBack。只有在业务方返回 RECONSUME_LATER 时，该批消息都需要发 ACK 消息，如果消息发送 ACK 失败，则直接将本批 ACK 消费发送失败的消息再次封装为 ConsumeRequest，然后延迟 5s 后重新消费。如果 ACK 消息发送成功，则该消息会延迟消费。

代码清单 5-57　ConsumeMessageConcurrentlyService#processConsumeResult

```
long offset = consumeRequest.getProcessQueue().
            removeMessage(consumeRequest.getMsgs());
if (offset >= 0 && !consumeRequest.getProcessQueue().isDropped()) {
    this.defaultMQPushConsumerImpl.getOffsetStore().updateOffset(consumeReques
        t.getMessageQueue(), offset, true);
}
```

Step11：从 ProcessQueue 中移除这批消息，这里返回的偏移量是移除该批消息后最小的偏移量，然后用该偏移量更新消息消费进度，以便在消费者重启后能从上一次的消费进度开始消费，避免消息重复消费。值得重点注意的是当消息监听器返回 RECONSUME_LATER，消息消费进度也会向前推进，用 ProcessQueue 中最小的队列偏移量调用消息消费进度存储器 OffsetStore 更新消费进度，这是因为当返回 RECONSUME_LATER，RocketMQ 会创建一条与原先消息属性相同的消息，拥有一个唯一的新 msgId，并存储原消息 ID，该消息会存入到 commitlog 文件中，与原先的消息没有任何关联，那该消息当然也会进入到 ConsuemeQueue 队列中，将拥有一个全新的队列偏移量。

并发消息消费的整体流程就介绍到这里，下文会对消息消费的其中两个重要步骤进行详细分析，ACK 消息发送与消息消费进度存储。

5.6.2　消息确认 (ACK)

如果消息监听器返回的消费结果为 RECONSUME_LATER，则需要将这些消息发送给 Broker 延迟消息。如果发送 ACK 消息失败，将延迟 5s 后提交线程池进行消费。ACK 消息发送的网络客户端入口：MQClientAPIImpl#consumerSendMessageBack，命令编码：RequestCode.CONSUMER_SEND_MSG_BACK，协议头部如图 5-12 所示。

下面让我们一一来分析 ConsumerSendMsgBackRequestHeader 的核心属性。

1）offset：消息物理偏移量。
2）group：消费组名。
3）delayLevel：延迟级别，RcketMQ 不支持精确的定时消息调度，而是提供几个延时级别，MessageStoreConfig#messageDelayLevel = "1s 5s 10s 30s 1m 2m 3m 4m 5m 6m 7m 8m 9m 10m 20m 30m 1h 2h"，如果 delayLevel=1 表示延迟 5s,delayLevel=2 则表示延迟 10s。

4）originMsgId：消息 ID。
5）originTopic：消息主题。
6）maxReconsumeTimes：最大重新消费次数，默认为 16 次。

```
ConsumerSendMsgBackRequestHeader
-private Long offset
-private String group
-private Integer delayLevel
-private String originMsgId
-private String originTopic
-private Integer maxReconsumeTimes
```

图 5-12　ACK 消息请求头部类图

客户端以同步方式发送 RequestCode.CONSUMER_SEND 到服务端。服务端命令处理器：org.apache.rocketmq.broker.processor.SendMessageProcessor#consumerSendMsgBack。

Step1：获取消费组的订阅配置信息，如果配置信息为空返回配置组信息不存在错误，如果重试队列数量小于 1，则直接返回成功，说明该消费组不支持重试。消费组核心类图如图 5-13 所示。

```
SubscriptionGroupConfig
-private String groupName
-private boolean consumeEnable = true
-private boolean consumeFromMinEnable = true
-private boolean consumeBroadcastEnable = true
-private int retryQueueNums = 1
-private int retryMaxTimes = 16
-private long brokerId = MixAll.MASTER_ID
-private long whichBrokerWhenConsumeSlowly = 1
-private boolean notifyConsumerIdsChangedEnable = true
```

图 5-13　消息订阅组配置信息

下面让我们一一来介绍 SubscriptionGroupConfig 的核心属性。

1）String groupName：消费组名。

2）consumeEnable：是否可以消费，默认该值为 true，如果 consumeEnable=false，该消费组无法拉取消息，从而无法消费消息。

3）consumeFromMinEnable：默认为 true，是否允许从队列最小偏移量开始消费，目前未使用该参数。

4）consumeBroadcastEnable：默认为 true，设置该消费组是否能以广播模式消费，如果设置为 false，则表示只能以集群模式消费。

5）retryQueueNums：重试队列个数，默认为 1，每一个 Broker 上一个重试队列。

6）retryMaxTimes：消息最大重试次数，默认为 16。

7）brokerId：masterId。

8）whichBrokerWhenConsumeSlowly：如果消息堵塞（主），将转向该 brokerId 的服务器上拉取消息，默认为 1。

9）notifyConsumerIdsChangedEnable：当消费发送变化时是否立即进行消息队列重新负载。

消费组订阅信息配置信息存储在 Broker 的 ${ROCKET_HOME}/store/config/subscriptionGroup.json。默认情况下 BrokerConfig.autoCreateSubscriptionGroup 默认为 true，表示在第一次使用消费组配置信息时如果不存在，则使用上述默认值自动创建一个，如果为 false，则只能通过客户端命令 mqadmin updateSubGroup 创建后修改相关参数。

代码清单 5-58　SendMessageProcessor#consumerSendMsgBack

```
String newTopic = MixAll.getRetryTopic(requestHeader.getGroup());
int queueIdInt = Math.abs(this.random.nextInt() % 99999999) %
    subscriptionGroupConfig.getRetryQueueNums();
```

Step2：创建重试主题，重试主题名称：%RETRY%+ 消费组名称，并从重试队列中随机选择一个队列，并构建 TopicConfig 主题配置信息。

代码清单 5-59　SendMessageProcessor#consumerSendMsgBack

```
MessageExt msgExt = this.brokerController.getMessageStore()
        .lookMessageByOffset(requestHeader.getOffset());
if (null == msgExt) {
    response.setCode(ResponseCode.SYSTEM_ERROR);
    response.setRemark("look message by offset failed, " +
            requestHeader.getOffset());
    return response;
}
final String retryTopic =
        msgExt.getProperty(MessageConst.PROPERTY_RETRY_TOPIC);
if (null == retryTopic) {
    MessageAccessor.putProperty(msgExt, MessageConst.PROPERTY_RETRY_TOPIC,
            msgExt.getTopic());
}
msgExt.setWaitStoreMsgOK(false);
```

Step3：根据消息物理偏移量从 commitlog 文件中获取消息，同时将消息的主题存入属性中。

Step4：设置消息重试次数，如果消息已重试次数超过 maxReconsumeTimes，再次改变 newTopic 主题为 DLQ（"%DLQ%"），该主题的权限为只写，说明消息一旦进入到 DLQ 队列中，RocketMQ 将不负责再次调度进行消费了，需要人工干预。

代码清单 5-60　SendMessageProcessor#consumerSendMsgBack

```
MessageExtBrokerInner msgInner = new MessageExtBrokerInner();
msgInner.setTopic(newTopic);
```

```
msgInner.setBody(msgExt.getBody());
msgInner.setFlag(msgExt.getFlag());
MessageAccessor.setProperties(msgInner, msgExt.getProperties());
msgInner.setPropertiesString(MessageDecoder.messageProperties2String(msgExt.g
    etProperties()));
msgInner.setTagsCode(MessageExtBrokerInner.tagsString2tagsCode(null,
    msgExt.getTags()));
msgInner.setQueueId(queueIdInt);
msgInner.setSysFlag(msgExt.getSysFlag());
msgInner.setBornTimestamp(msgExt.getBornTimestamp());
msgInner.setBornHost(msgExt.getBornHost());
msgInner.setStoreHost(this.getStoreHost());
msgInner.setReconsumeTimes(msgExt.getReconsumeTimes() + 1);
String originMsgId = MessageAccessor.getOriginMessageId(msgExt);
MessageAccessor.setOriginMessageId(msgInner, UtilAll.isBlank(originMsgId) ?
    msgExt.getMsgId() : originMsgId);
```

Step5：根据原先的消息创建一个新的消息对象，重试消息会拥有自己的唯一消息 ID（msgId）并存入到 commitlog 文件中，并不会去更新原先消息，而是会将原先的主题、消息 ID 存入消息的属性中，主题名称为重试主题，其他属性与原先消息保持相同。

Step6：将消息存入到 CommitLog 文件中。该部分逻辑在前面章节中已详细介绍，这里想再重点突出一个机制，消息重试机制依托于定时任务实现，具体如下。

代码清单 5-61　CommitLog#putMessage

```
// Delay Delivery
if (msg.getDelayTimeLevel() > 0) {
    if (msg.getDelayTimeLevel() >
            this.defaultMessageStore.getScheduleMessageService()
                .getMaxDelayLevel()) {

        msg.setDelayTimeLevel(this.defaultMessageStore.
            getScheduleMessageService().getMaxDelayLevel());
    }
    topic = ScheduleMessageService.SCHEDULE_TOPIC;
    queueId = ScheduleMessageService.delayLevel2QueueId
        (msg.getDelayTimeLevel());
    // Backup real topic, queueId
    MessageAccessor.putProperty(msg, MessageConst.PROPERTY_REAL_TOPIC,
        msg.getTopic());
    MessageAccessor.putProperty(msg, MessageConst.PROPERTY_REAL_QUEUE_ID,
        String.valueOf(msg.getQueueId()));
    msg.setPropertiesString(MessageDecoder.messageProperties2String(msg.getPro
        perties()));
    msg.setTopic(topic);
    msg.setQueueId(queueId);
}
```

在存入 Commitlog 文件之前，如果消息的延迟级别 delayTimeLevel 大于 0，替换消息的主题与队列为定时任务主题"SCHEDULE_TOPIC_XXXX",队列 ID 为延迟级别减 1。再次将消息主题、队列存入消息的属性中,键分别为:PROPERTY_REAL_TOPIC、PROPERTY_REAL_QUEUE_ID。

ACK 消息存入 CommitLog 文件后,将依托 RocketMQ 定时消息机制在延迟时间到期后再次将消息拉取,提交消费线程池,有关定时任务机制将在 5.7 节详细分析。ACK 消息是同步发送的,如果在发送过程中出现错误,将记录所有发送 ACK 消息失败的消息,然后再次封装成 ConsumeRequest,延迟 5s 执行。

5.6.3 消费进度管理

消息消费者在消费一批消息后,需要记录该批消息已经消费完毕,否则当消费者重新启动时又得从消息消费队列的开始消费,这显然是不能接受的。从 5.6.1 节也可以看到,一次消息消费后会从 ProceeQueue 处理队列中移除该批消息,返回 ProceeQueue 最小偏移量,并存入消息进度表中。那消息进度文件存储在哪合适呢?

广播模式:同一个消费组的所有消息消费者都需要消费主题下的所有消息,也就是同组内的消费者的消息消费行为是对立的,互相不影响,故消息进度需要独立存储,最理想的存储地方应该是与消费者绑定。

集群模式:同一个消费组内的所有消息消费者共享消息主题下的所有消息,同一条消息(同一个消息消费队列)在同一时间只会被消费组内的一个消费者消费,并且随着消费队列的动态变化重新负载,所以消费进度需要保存在一个每个消费者都能访问到的地方。

RocketMQ 消息消费进度接口如图 5-14 所示。

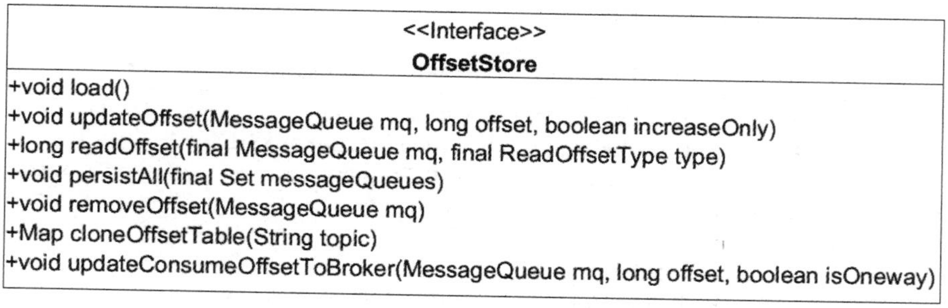

```
          <<Interface>>
           OffsetStore
+void load()
+void updateOffset(MessageQueue mq, long offset, boolean increaseOnly)
+long readOffset(final MessageQueue mq, final ReadOffsetType type)
+void persistAll(final Set messageQueues)
+void removeOffset(MessageQueue mq)
+Map cloneOffsetTable(String topic)
+void updateConsumeOffsetToBroker(MessageQueue mq, long offset, boolean isOneway)
```

图 5-14 消息进度 OffsetStore 类图

1)void load()

从消息进度存储文件加载消息进度到内存。

2)void updateOffset(final MessageQueue mq, final long offset, final boolean increaseOnly)

更新内存中的消息消费进度。

MessageQueue mq:消息消费队列。

long offset：消息消费偏移量。

increaseOnly：true 表示 offset 必须大于内存中当前的消费偏移量才更新。

3）long readOffset（final MessageQueue mq, final ReadOffsetType type）

读取消息消费进度。

mq：消息消费队列。

ReadOffsetType type：读取方式，可选值 READ_FROM_MEMORY：从内存中；READ_FROM_STORE：从磁盘中；MEMORY_FIRST_THEN_STORE：先从内存读取，再从磁盘。

4）void persistAll（final Set<MessageQueue> mqs）

持久化指定消息队列进度到磁盘。

Set<MessageQueue> mqs：消息队列集合。

5）void removeOffset（MessageQueue mq）

将消息队列的消息消费进度从内存中移除。

6）Map<MessageQueue, Long> cloneOffsetTable（String topic）

克隆该主题下所有消息队列的消息消费进度。

7）void updateConsumeOffsetToBroker（MessageQueue mq, long offset, boolean isOneway）

更新存储在 Brokder 端的消息消费进度，使用集群模式。

1. 广播模式消费进度存储

广播模式消息消费进度存储在消费者本地，其实现类 org.apache.rocketmq.client.consumer.store.LocalFileOffsetStore。

代码清单 5-62　LocalFileOffsetStore

```
public final static String LOCAL_OFFSET_STORE_DIR =
    System.getProperty("rocketmq.client.localOffsetStoreDir",
        System.getProperty("user.home") + File.separator + ".rocketmq_offsets");
private final MQClientInstance mQClientFactory;
private final String groupName;
private final String storePath;
private ConcurrentMap<MessageQueue, AtomicLong> offsetTable =
    new ConcurrentHashMap<MessageQueue, AtomicLong>();
```

1）LOCAL_OFFSET_STORE_DIR，消息进度存储目录，可以通过 -Drocketmq.client.localOffsetStoreDir，如果未指定，则默认为用户主目录 /.rocketmq_offsets。

2）MQClientInstance mQClientFactory：消息客户端。

3）String groupName：消息消费组。

4）String storePath：消息进度存储文件，LOCAL_OFFSET_STORE_DIR/.rocketmq_offsets/{mQClientFactory.getClientId()}/groupName/offsets.json。

5）ConcurrentMap<MessageQueue, AtomicLong> offsetTable：消息消费进度（内存）。
下面对 LocalFileOffsetStore 核心方法做个简单介绍。
❏ load 方法

代码清单 5-63　LocalFileOffsetStore#load

```
public void load() throws MQClientException {
    OffsetSerializeWrapper offsetSerializeWrapper = this.readLocalOffset();
    if (offsetSerializeWrapper != null &&
            offsetSerializeWrapper.getOffsetTable() != null) {
        offsetTable.putAll(offsetSerializeWrapper.getOffsetTable());
        for (MessageQueue mq : offsetSerializeWrapper.getOffsetTable().keySet()) {
            AtomicLong offset = offsetSerializeWrapper.getOffsetTable().get(mq);
            log.info("load consumer's offset, {} {} {}",this.groupName,mq,
                    offset.get());
        }
    }
}
```

首先看一下 OffsetSerializeWrapper 内部就是 ConcurrentMap<MessageQueue, AtomicLong> offsetTable 数据结构的封装，readLocakOffset 方法首先从 storePath 中尝试加载，如果从该文件读取到内容为空，尝试从 storePath+".bak" 中尝试加载，如果还是未找到，则返回 null。为了对消息进度有一个更直观的了解，消息进度文件存储内容如图 5-15 所示。

```
{
    "offsetTable":[{
        "brokerName":"broker-a",
        "queueId":3,
        "topic":"TopicTest"
    }:2,{
        "brokerName":"broker-a",
        "queueId":2,
        "topic":"TopicTest"
    }:1,{
        "brokerName":"broker-a",
        "queueId":1,
        "topic":"TopicTest"
    }:2,{
        "brokerName":"broker-a",
        "queueId":0,
        "topic":"TopicTest"
    }:1
    }
}
```

图 5-15　消息进度文件内容

广播模式消费进度与消费组没啥关系，直接保存 MessageQueue：Offset。
❏ persistAll(Set<MessageQueue> mqs) 持久化消息进度

代码清单 5-64　LocalFileOffsetStore#persistAll

```java
public void persistAll(Set<MessageQueue> mqs) {
    if (null == mqs || mqs.isEmpty())
        return;
    OffsetSerializeWrapper offsetSerializeWrapper = new OffsetSerializeWrapper();
    for (Map.Entry<MessageQueue, AtomicLong> entry :
            this.offsetTable.entrySet()) {
        if (mqs.contains(entry.getKey())) {
            AtomicLong offset = entry.getValue();
            offsetSerializeWrapper.getOffsetTable().put(entry.getKey(), offset);
        }
    }
    String jsonString = offsetSerializeWrapper.toJson(true);
    if (jsonString != null) {
        try {
            MixAll.string2File(jsonString, this.storePath);
        } catch (IOException e) {
            log.error("persistAll consumer offset Exception, " + this.storePath, e);
        }
    }
}
```

持久化消息进度就是将 ConcurrentMap<MessageQueue, AtomicLong> offsetTable 序列化到磁盘文件中。代码不容易理解，关键是什么时候持久化消息消费进度。原来在 MQClientInstance 中会启动一个定时任务，默认每 5s 持久化一次，可通过 persistConsumer-OffsetInterval 设置。

代码清单 5-65　LocalFileOffsetStore#persistAll

```java
this.scheduledExecutorService.scheduleAtFixedRate(new Runnable() {
    public void run() {
        try {
            MQClientInstance.this.persistAllConsumerOffset();
        } catch (Exception e) {
            log.error("ScheduledTask persistAllConsumerOffset exception", e);
        }
    }
}, 1000 * 10, this.clientConfig.getPersistConsumerOffsetInterval(),
        TimeUnit.MILLISECONDS);
```

广播模式消息消费进度的存储、更新、持久化还是比较容易的，本书就简单介绍到这里，接下来重点分析集群模式消息进度管理。

2. 集群模式消费进度存储

集群模式消息进度存储文件存放在消息服务端 Broker。消息消费进度集群模式实现类：org.apache.rocketmq.client.consumer.store.RemoteBrokerOffsetStore。其整个实现原理如图 5-16 所示。

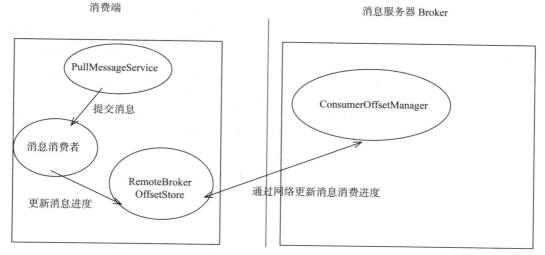

图 5-16　集群模式消息消费进度实现原理图

消息消费进度的读取，持久化与广播模式实现细节差不多，集群模式消息消费进度如果从内存中读取消费进度，则从 RemoteBrokerOffsetStore 的 ConcurrentMap<MessageQueue, AtomicLong> offsetTable =new ConcurrentHashMap<MessageQueue, AtomicLong>() 中根据消息消费队列获取其消息消费进度；如果从磁盘读取，则发送网络请求，请求命令为 QUERY_CONSUMER_OFFSET。持久化消息进度，则请求命令为 UPDATE_CONSUMER_OFFSET，更新 ConsumerOffsetManager 的 ConcurrentMap<String/* topic@group */, ConcurrentMap<Integer/* 消息队列 ID*/, Long/* 消息消费进度 */>> offsetTable，Broker 端默认 10s 持久化一次消息进度，存储文件名：${RocketMQ_HOME}/store/config/consumerOffset.json。请存储内容如图 5-17 所示。

```
{
    "offsetTable":{
        "TopicTest@DataSyncConnsumeGroup":{0:38,2:37,1:37,3:38
        },
        "%RETRY%DataSyncConnsumeGroup@DataSyncConnsumeGroup":{0:0
        }
    }
}
```

图 5-17　集群模式消息消费进度存储

3. 消费进度设计思考

消息消费进度的存储，广播模式与消费组无关，集群模式下以主题与消费组为键保存该主题所有队列的消费进度。结合并发消息消费的整个流程，思考一下并发消息消费关于消息进度更新的问题，顺序消息消费将在 5.9 节中重点讨论。

1）消费者线程池每处理完一个消息消费任务（ConsumeRequest）时会从 ProceeQueue

中移除本批消费的消息，并返回 ProceeQueue 中最小的偏移量，用该偏移量更新消息队列消费进度，也就是说更新消费进度与消费任务中的消息没什么关系。例如现在两个消费任务 task1（queueOffset 分别为 20,40），task2（50,70），并且 ProceeQueue 中当前包含最小消息偏移量为 10 的消息，则 task2 消费结束后，将使用 10 去更新消费进度，并不会是 70。当 task1 消费结束后，还是以 10 去更新消息队列消费进度，消息消费进度的推进取决于 ProceeQueue 中偏移量最小的消息消费速度。如果偏移量为 10 的消息消费成功后，假如 ProceeQueue 中包含消息偏移量为 100 的消息，则消息偏移量为 10 的消息消费成功后，将直接用 100 更新消息消费进度。那如果在消费消息偏移量为 10 的消息时发送了死锁导致一直无法被消费，那岂不是消息进度无法向前推进。是的，为了避免这种情况，RocketMQ 引入了一种消息拉取流控措施：DefaultMQPushConsumer#consumeConcurrentlyMaxSpan=2000，消息处理队列 ProceeQueue 中最大消息偏移与最小偏移量不能超过该值，如超过该值，触发流控，将延迟该消息队列的消息拉取。

2）触发消息消费进度更新的另外一个是在进行消息负载时，如果消息消费队列被分配给其他消费者时，此时会将该 ProceeQueue 状态设置为 droped，持久化该消息队列的消费进度，并从内存中移除。

5.7 定时消息机制

定时消息是指消息发送到 Broker 后，并不立即被消费者消费而是要等到特定的时间后才能被消费，RocketMQ 并不支持任意的时间精度，如果要支持任意时间精度的定时调度，不可避免地需要在 Broker 层做消息排序（可以参考 JDK 并发包调度线程池 ScheduledExecutorService 的实现原理），再加上持久化方面的考量，将不可避免地带来具大的性能消耗，所以 RocketMQ 只支持特定级别的延迟消息。消息延迟级别在 Broker 端通过 messageDelayLevel 配置，默认为 "1s 5s 10s 30s 1m 2m 3m 4m 5m 6m 7m 8m 9m 10m 20m 30m 1h 2h"，delayLevel=1 表示延迟 1s,delayLevel=2 表示延迟 5s,依次类推。说到定时任务，上文提到的消息重试正是借助定时任务实现的，在将消息存入 commitlog 文件之前需要判断消息的重试次数，如果大于 0，则会将消息的主题设置为 SCHEDULE_TOPIC_XXXX。RocketMQ 定时消息实现类为 org.apache.rocketmq.store.schedule.ScheduleMessageService。该类的实例在 DefaultMessageStore 中创建，通过在 DefaultMessageStore 中调用 load 方法加载并调用 start 方法进行启动。接下来我们分析一下 ScheduleMessageService 实现原理。

ScheduleMessageService 如图 5-18 所示。

下面让我们一一来分析 ScheduleMessageService 的核心属性。

1）SCHEDULE_TOPIC：定时消息统一主题。

2）FIRST_DELAY_TIME：第一次调度时延迟的时间，默认为 1s。

3）DELAY_FOR_A_WHILE：每一延时级别调度一次后延迟该时间间隔后再放入调度池。

```
                    ScheduleMessageService
-public static final String SCHEDULE_TOPIC = "SCHEDULE_TOPIC_XXXX"
-private static final long FIRST_DELAY_TIME = 1000L
-private static final long DELAY_FOR_A_WHILE = 100L
-private static final long DELAY_FOR_A_PERIOD = 10000L
-private final ConcurrentMap delayLevelTable
-private final ConcurrentMap offsetTable
-private final Timer timer = new Timer("ScheduleMessageTimerThread", true)
-private final DefaultMessageStore defaultMessageStore
-private int maxDelayLevel
```

图 5-18　ScheduleMessageService 类图

4）DELAY_FOR_A_PERIOD：发送异常后延迟该时间后再继续参与调度。

5）ConcurrentMap<Integer /* level */, Long/* delay timeMillis */> delayLevelTable：延迟级别，将 "1s 5s 10s 30s 1m 2m 3m 4m 5m 6m 7m 8m 9m 10m 20m 30m 1h 2h" 字符串解析成 delayLevelTable，转换后的数据结构类似 {1:1000,2:5000,3:30000,...}。

6）ConcurrentMap<Integer /* level */, Long/* offset */> offsetTable：延迟级别消息消费进度。

7）DefaultMessageStore defaultMessageStore：默认消息存储器。

8）int maxDelayLevel：MessageStoreConfig#messageDelayLevel 中最大消息延迟级别。

ScheduleMessageService 方法的调用顺序：构造方法 ->load()->start() 方法。

5.7.1　load 方法

ScheduleMessageService 继承自 ConfigManager，load 方法的代码如下。

代码清单 5-66　ScheduleMessageService#load

```
public boolean load() {
    boolean result = super.load();
    result = result && this.parseDelayLevel();
    return result;
}
```

该方法主要完成延迟消息消费队列消息进度的加载与 delayLevelTable 数据的构造，延迟队列消息消费进度默认存储路径为 ${ROCKET_HOME}/store/config/delayOffset.json，存储格式为如图 5-19 所示。

```
{
    "offsetTable":{12:0,6:0,13:0,5:1,18:0,7:0,8:0,17:0,9:0,10:0,16:0,15:0,14:0,3:22,11:0,4:1
    }
}
```

图 5-19　延迟队列消息消费进度

同时解析 MessageStoreConfig#messageDelayLevel 定义的延迟级别转换为 Map，延迟级别 1,2,3 等对应的延迟时间。

5.7.2 start 方法

start 根据延迟级别创建对应的定时任务，启动定时任务持久化延迟消息队列进度存储。

代码清单 5-67　ScheduleMessageService#start

```
for (Map.Entry<Integer, Long> entry : this.delayLevelTable.entrySet()) {
    Integer level = entry.getKey();
    Long timeDelay = entry.getValue();
    Long offset = this.offsetTable.get(level);
    if (null == offset) {
        offset = 0L;
    }
    if (timeDelay != null) {
        this.timer.schedule(new DeliverDelayedMessageTimerTask(level, offset),
            FIRST_DELAY_TIME);
    }
}
```

Step1：根据延迟队列创建定时任务，遍历延迟级别，根据延迟级别 level 从 offsetTable 中获取消费队列的消费进度，如果不存在，则使用 0。也就是说每一个延迟级别对应一个消息消费队列。然后创建定时任务，每一个定时任务第一次启动时默认延迟 1s 先执行一次定时任务，第二次调度开始才使用相应的延迟时间。延迟级别与消息消费队列的映射关系为：消息队列 ID= 延迟级别 −1。

代码清单 5-68　ScheduleMessageService#queueId2DelayLevel

```
public static int queueId2DelayLevel(final int queueId) {
    return queueId + 1;
}
public static int delayLevel2QueueId(final int delayLevel) {
    return delayLevel - 1;
}
```

定时消息的第一个设计关键点是，定时消息单独一个主题：SCHEDULE_TOPIC_XXXX，该主题下队列数量等于 MessageStoreConfig#messageDelayLevel 配置的延迟级别数量，其对应关系为 queueId 等于延迟级别减 1。ScheduleMessageService 为每一个延迟级别创建一个定时 Timer 根据延迟级别对应的延迟时间进行延迟调度。在消息发送时，如果消息的延迟级别 delayLevel 大于 0，将消息的原主题名称、队列 ID 存入消息的属性中，然后改变消息的主题、队列与延迟主题与延迟主题所属队列，消息将最终转发到延迟队列的消费队列。

代码清单 5-69　ScheduleMessageService#start

```
this.timer.scheduleAtFixedRate(new TimerTask() {
    public void run() {
```

```
            try {
                ScheduleMessageService.this.persist();
            } catch (Throwable e) {
                log.error("scheduleAtFixedRate flush exception", e);
            }
        }
    }, 10000, this.defaultMessageStore.getMessageStoreConfig()
            .getFlushDelayOffsetInterval());
```

Step2：创建定时任务，每隔 10s 持久化一次延迟队列的消息消费进度（延迟消息调进度），持久化频率可以通过 flushDelayOffsetInterval 配置属性进行设置。

5.7.3 定时调度逻辑

ScheduleMessageService 的 start 方法启动后，会为每一个延迟级别创建一个调度任务，每一个延迟级别其实对应 SCHEDULE_TOPIC_XXXX 主题下的一个消息消费队列。定时调度任务的实现类为 DeliverDelayedMessageTimerTask，其核心实现为 executeOnTimeup。

代码清单 5-70 ScheduleMessageService$DeliverDelayedMessageTimerTask#executeOnTimeup

```
ConsumeQueue cq = ScheduleMessageService.this.defaultMessageStore.
        findConsumeQueue(SCHEDULE_TOPIC, delayLevel2QueueId(delayLevel));
```

Step1：根据队列 ID 与延迟主题查找消息消费队列，如果未找到，说明目前并不存在该延时级别的消息，忽略本次任务，根据延时级别创建下一次调度任务即可。

代码清单 5-71 ScheduleMessageService$DeliverDelayedMessageTimerTask#executeOnTimeup

```
SelectMappedBufferResult bufferCQ = cq.getIndexBuffer(this.offset);
```

Step2：根据 offset 从消息消费队列中获取当前队列中所有有效的消息。如果未找到，则更新一下延迟队列定时拉取进度并创建定时任务待下一次继续尝试。

代码清单 5-72 ScheduleMessageService$DeliverDelayedMessageTimerTask#executeOnTimeup

```
long nextOffset = offset;
int i = 0;
ConsumeQueueExt.CqExtUnit cqExtUnit = new ConsumeQueueExt.CqExtUnit();
for (; i < bufferCQ.getSize(); i += ConsumeQueue.CQ_STORE_UNIT_SIZE) {
    long offsetPy = bufferCQ.getByteBuffer().getLong();
    int sizePy = bufferCQ.getByteBuffer().getInt();
    long tagsCode = bufferCQ.getByteBuffer().getLong();
    long now = System.currentTimeMillis();
    long deliverTimestamp = this.correctDeliverTimestamp(now, tagsCode);
    nextOffset = offset + (i / ConsumeQueue.CQ_STORE_UNIT_SIZE);
    // .....
}
```

Step3：遍历 ConsumeQueue，每一个标准 ConsumeQueue 条目为 20 个字节。解析出消

息的物理偏移量、消息长度、消息 tag hashcode，为从 commitlog 加载具体的消息做准备。

代码清单 5-73　ScheduleMessageService$DeliverDelayedMessageTimerTask#executeOnTimeup

```
MessageExt msgExt = ScheduleMessageService.this
    .defaultMessageStore.lookMessageByOffset(offsetPy, sizePy);
```

Step4：根据消息物理偏移量与消息大小从 commitlog 文件中查找消息。如果未找到消息，打印错误日志，根据延迟时间创建下一个定时器。

代码清单 5-74　ScheduleMessageService$DeliverDelayedMessageTimerTask#messageTimeup

```
msgInner.setReconsumeTimes(msgExt.getReconsumeTimes());
msgInner.setWaitStoreMsgOK(false);
MessageAccessor.clearProperty(msgInner,
    MessageConst.PROPERTY_DELAY_TIME_LEVEL);
msgInner.setTopic(msgInner.getProperty(MessageConst.PROPERTY_REAL_TOPIC));
String queueIdStr = msgInner.getProperty(MessageConst.PROPERTY_REAL_QUEUE_ID);
int queueId = Integer.parseInt(queueIdStr);
msgInner.setQueueId(queueId);
```

Step5：根据消息重新构建新的消息对象，清除消息的延迟级别属性（delayLevel）、并恢复消息原先的消息主题与消息消费队列，消息的消费次数 reconsumeTimes 并不会丢失。

代码清单 5-75　ScheduleMessageService$DeliverDelayedMessageTimerTask#executeOnTimeup

```
PutMessageResult putMessageResult = ScheduleMessageService.this
    .defaultMessageStore.putMessage(msgInner);
```

Step6：将消息再次存入到 commitlog，并转发到主题对应的消息队列上，供消费者再次消费。

Step7：更新延迟队列拉取进度。

定时消息的第二个设计关键点：消息存储时如果消息的延迟级别属性 delayLevel 大于 0，则会备份原主题、原队列到消息属性中，其键分别为 PROPERTY_REAL_TOPIC、PROPERTY_REAL_QUEUE_ID，通过为不同的延迟级别创建不同的调度任务，当时间到达后执行调度任务，调度任务主要就是根据延迟拉取消息消费进度从延迟队列中拉取消息，然后从 commitlog 中加载完整消息，清除延迟级别属性并恢复原先的主题、队列，再次创建一条新的消息存入到 commitlog 中并转发到消息消费队列供消息消费者消费。

上述就是定时消息的实现原理，其整个流程如图 5-20 所示。

1）消息消费者发送消息，如果发送消息的 delayLevel 大于 0，则改变消息主题为 SCHEDULE_TOPIC_XXXX，消息队列为 delayLevel 减 1。

2）消息经由 commitlog 转发到消息消费队列 SCHEDULE_TOPIC_XXXX 的消息消费队列 0。

3）定时任务 Time 每隔 1s 根据上次拉取偏移量从消费队列中取出所有消息。

4）根据消息的物理偏移量与消息大小从 CommitLog 中拉取消息。

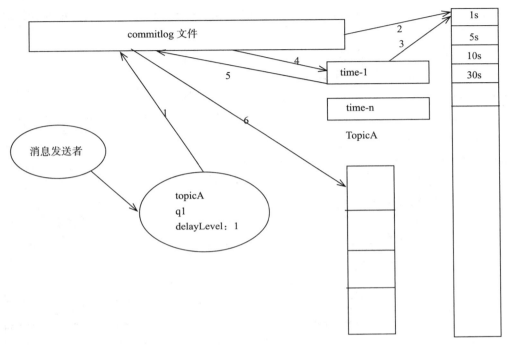

图 5-20　定时消息实现流程图

5）根据消息属性重新创建消息，并恢复原主题 topicA、原队列 ID，清除 delayLevel 属性，存入 commitlog 文件。

6）转发到原主题 topicA 的消息消费队列，供消息消费者消费。

5.8　消息过滤机制

RocketMQ 支持表达式过滤与类过滤两种模式，其中表达式又分为 TAG 和 SQL92。类过滤模式允许提交一个过滤类到 FilterServer，消息消费者从 FilterServer 拉取消息，消息经过 FilterServer 时会执行过滤逻辑。表达式模式分为 TAG 与 SQL92 表达式，SQL92 表达式以消息属性过滤上下文，实现 SQL 条件过滤表达式而 TAG 模式就是简单为消息定义标签，根据消息属性 tag 进行匹配。消息过滤 API 如图 5-21 所示。

<<Interface>> **MessageFilter**
+boolean isMatchedByConsumeQueue(Long tagsCode, CqExtUnit cqExtUnit)
+boolean isMatchedByCommitLog(final ByteBuffer msgBuffer, final Map properties)

图 5-21　MessageFilter 类图

下面让我们一一来分析 MessageFilter 的核心接口。

1）boolean isMatchedByConsumeQueue（final Long tagsCode,final ConsumeQueueExt.CqExtUnit cqExtUnit）

根据 ConsumeQueue 判断消息是否匹配。

Long tagsCode：消息 tag 的 hashcode。

ConsumeQueueExt.CqExtUnit：consumequeue 条目扩展属性。

2）boolean isMatchedByCommitLog（final ByteBuffer msgBuffer, final Map<String,String> properties）

根据存储在 commitlog 文件中的内容判断消息是否匹配。

ByteBuffer msgBuffer：消息内容，如果为空，该方法返回 true。

Map<String,String> properties：消息属性，主要用于表达式 SQL92 过滤模式。

本节重点探讨 RocketMQ 基于表达式的消息过滤机制，基于类模式的消息过滤将在第 6 章详细描述。RocketMQ 消息过滤方式不同于其他消息中间件，是在订阅时做过滤，从第 4 章的介绍中我们知道 ConsumeQueue 的存储格式如图 5-22 所示。

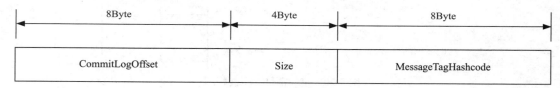

图 5-22　ConsumeQueue 存储格式

消息发送者在消息发送时如果设置了消息的 tags 属性，存储在消息属性中，先存储在 CommitLog 文件中，然后转发到消息消费队列，消息消费队列会用 8 个字节存储消息 tag 的 hashcode，之所以不直接存储 tag 字符串，是因为将 ConumeQueue 设计为定长结构，加快消息消费的加载性能。在 Broker 端拉取消息时，遍历 ConsumeQueue，只对比消息 tag 的 hashcode，如果匹配则返回，否则忽略该消息。Consume 在收到消息后，同样需要先对消息进行过滤，只是此时比较的是消息 tag 的值而不再是 hashcode。

接下来从源码的角度探究 RocketMQ 是如何实现的。

代码清单 5-76　DefaultMQPushConsumerImpl#subscribe

```
public void subscribe(String topic, String subExpression) throws 
    MQClientException {
    try {
        SubscriptionData subscriptionData = FilterAPI.buildSubscriptionData(
            this.defaultMQPushConsumer.getConsumerGroup(),topic, subExpression);
        this.rebalanceImpl.getSubscriptionInner().put(topic, subscriptionData);
        if (this.mQClientFactory != null) {
            this.mQClientFactory.sendHeartbeatToAllBrokerWithLock();
        }
```

```
        } catch (Exception e) {
            throw new MQClientException("subscription exception", e);
        }
    }
```

Step1：消费者订阅消息主题与消息过滤表达式。构建订阅信息并加入到 RebalanceImpl 中，以便 RebalanceImpl 进行消息队列负载，订阅过滤数据类图如图 5-23 所示。

SubscriptionData
-public final static String SUB_ALL = "*"
-private boolean classFilterMode = false
-private String topic
-private String subString
-private Set tagsSet = new HashSet()
-private Set codeSet = new HashSet()
-private String expressionType

ExpressionType
-public static final String SQL92 = "SQL92"
-public static final String TAG = "TAG"

图 5-23　消息订阅消息

下面让我们一一来介绍 SubscriptionData 的核心属性。

1）String SUB_ALL：过滤模式，默认为全匹配。
2）boolean classFilterMode：是否是类过滤模式，默认为 false。
3）String topic：消息主题名称。
4）String subString：消息过滤表达式，多个用双竖线隔开，例如"TAGA|| TAGB"。
5）Set<String> tagsSet：消息过滤 tag 集合，消费端过滤时进行消息过滤的依据。
6）Set<String> codeSet：消息过滤 tag hashcode 集合。
7）String expressionType：过滤类型，TAG 或 SQL92。

代码清单 5-77　DefaultMQPushConsumerImpl#pullMessage

```
String subExpression = null;
boolean classFilter = false;
SubscriptionData sd = this.rebalanceImpl.getSubscriptionInner().
            get(pullRequest.getMessageQueue().getTopic());
if (sd != null) {
    if (this.defaultMQPushConsumer.isPostSubscriptionWhenPull()
            && !sd.isClassFilterMode()) {
        subExpression = sd.getSubString();
    }
    classFilter = sd.isClassFilterMode();
}
int sysFlag = PullSysFlag.buildSysFlag(
        commitOffsetEnable, // commitOffset
        true, // suspend
        subExpression != null, // subscription
        classFilter // class filter
);
```

Step2：根据订阅消息构建消息拉取标记，设置 subExpression、classFilter 等与消息过滤相关。

代码清单 5-78　PullMessageProcessor#processRequest

```
subscriptionData = FilterAPI.build(
    requestHeader.getTopic(), requestHeader.getSubscription(),
    requestHeader.getExpressionType()
);
if (!ExpressionType.isTagType(subscriptionData.getExpressionType())) {
    consumerFilterData = ConsumerFilterManager.build(
        requestHeader.getTopic(), requestHeader.getConsumerGroup(),
        requestHeader.getSubscription(),requestHeader.getExpressionType(),
        requestHeader.getSubVersion()
    );
    assert consumerFilterData != null;
}
```

Step3：根据主题、消息过滤表达式构建订阅消息实体。如果是不是 TAG 模式，构建过滤数据 ConsumeFilterData。

代码清单 5-79　PullMessageProcessor#processRequest

```
MessageFilter messageFilter;
if (this.brokerController.getBrokerConfig().isFilterSupportRetry()) {
    messageFilter = new ExpressionForRetryMessageFilter(subscriptionData,
        consumerFilterData,this.brokerController.getConsumerFilterManager());
} else {
    messageFilter = new ExpressionMessageFilter(subscriptionData,
        consumerFilterData,this.brokerController.getConsumerFilterManager());
}
```

Step4：构建消息过滤对象，ExpressionForRetryMessageFilter，支持对重试主题的过滤，ExpressionMessageFilter，不支持对重试主题的属性过滤，也就是如果是 tag 模式，执行 isMatchedByCommitLog 方法将直接返回 true。

代码清单 5-80　DefaultMessageStore#getMessage

```
if (messageFilter != null && !messageFilter.isMatchedByConsumeQueue(tagsCode,
        extRet ? cqExtUnit : null)) {
    if (getResult.getBufferTotalSize() == 0) {
        status = GetMessageStatus.NO_MATCHED_MESSAGE;
    }
    continue;
}
```

Step5：根据偏移量拉取消息后，首先根据 ConsumeQueue 条目进行消息过滤，如果不匹配则直接跳过该条消息，继续拉取下一条消息。

代码清单 5-81　DefaultMessageStore#getMessage

```
if (messageFilter != null
        && !messageFilter.isMatchedByCommitLog(selectResult.getByteBuffer().
            slice(), null)) {
    if (getResult.getBufferTotalSize() == 0) {
        status = GetMessageStatus.NO_MATCHED_MESSAGE;
    }
    // release...
    selectResult.release();
    continue;
}
```

Step6：如果消息根据 ConsumeQueue 条目通过过滤，则需要从 CommitLog 文件中加载整个消息体，然后根据属性进行过滤。当然如果过滤方式是 TAG 模式，该方法默认返回 true，下文会对该方法详细讲解。

至此消息在消费拉取服务端的消息过滤流程就基本结束了，RocketMQ 会在消息接收端再次进行消息过滤。在讲解消费端消息过滤之前，先以 ExpressionMessageFilter 为例分析一下消息过滤实现细节。

代码清单 5-82　ExpressionMessageFilter#isMatchedByConsumeQueue

```
if (null == subscriptionData) {
    return true;
}
if (subscriptionData.isClassFilterMode()) {
    return true;
}
// by tags code.
if (ExpressionType.isTagType(subscriptionData.getExpressionType())) {
    if (tagsCode == null || tagsCode < 0L) {
        return true;
    }
    if (subscriptionData.getSubString().equals(SubscriptionData.SUB_ALL)) {
        return true;
    }
    return subscriptionData.getCodeSet().contains(tagsCode.intValue());
}
```

如果订阅消息为空，返回 true，不过滤；如果是类过滤模式，返回 true；如果是 TAG 过滤模式，并且消息的 tagsCode 为空或 tagsCode 小于 0，返回 true，说明消息在发送时没有设置 tag。如果订阅消息的 TAG hashcodes 集合中包含消息的 tagsCode，返回 true。基于 TAG 模式，根据 ConsumeQueue 进行消息过滤时只对比 tag 的 hashcode，所以基于 TAG 模式消息过滤，还需要在消息消费端对消息 tag 进行精确匹配。

代码清单 5-83　ExpressionMessageFilter#isMatchedByCommitLog

```
if (subscriptionData == null) {
    return true;
```

```
}
if (subscriptionData.isClassFilterMode()) {
    return true;
}
if (ExpressionType.isTagType(subscriptionData.getExpressionType())) {
    return true;
}
```

如果订阅信息为空,返回 true;如果是类过滤模式,返回 true;如果是 TAG 模式,返回 true。该方法主要是为表达式模式 SQL92 服务的,根据消息属性实现类似于数据库 SQL where 条件过滤方式。本书将不针对 SQL92 模式消息过滤详细讲解,但在第 9 章会给出 SQL92 过滤实例。

代码清单 5-84　PullAPIWrapper#processPullResult

```
if (PullStatus.FOUND == pullResult.getPullStatus()) {
    ByteBuffer byteBuffer = ByteBuffer.wrap(pullResultExt.getMessageBinary());
    List<MessageExt> msgList = MessageDecoder.decodes(byteBuffer);
    List<MessageExt> msgListFilterAgain = msgList;
    if (!subscriptionData.getTagsSet().isEmpty()
            && !subscriptionData.isClassFilterMode()) {
        msgListFilterAgain = new ArrayList<MessageExt>(msgList.size());
        for (MessageExt msg : msgList) {
            if (msg.getTags() != null) {
                if (subscriptionData.getTagsSet().contains(msg.getTags())) {
                    msgListFilterAgain.add(msg);
                }
            }
        }
    }
}
```

从消息拉取流程知道,消息拉取线程 PullMessageService 默认会使用异步方式从服务器拉取消息,消息消费端会通过 PullAPIWrapper 从响应结果解析出拉取到的消息。如果消息过滤模式为 TAG 模式,并且订阅 TAG 集合不为空,则对消息的 tag 进行判断,如果集合中包含消息的 TAG 则返回给消费者消费,否则跳过。

5.9　顺序消息

RocketMQ 支持局部消息顺序消费,可以确保同一个消息消费队列中的消息被顺序消费,如果需要做到全局顺序消费则可以将主题配置成一个队列,例如数据库 BinLog 等要求严格顺序的场景。根据并发消息消费的流程,消息消费包含如下 4 个步骤:消息队列负载、消息拉取、消息消费、消息消费进度存储。

5.9.1 消息队列负载

RocketMQ 首先需要通过 RebalanceService 线程实现消息队列的负载，集群模式下同一个消费组内的消费者共同承担其订阅主题下消息队列的消费，同一个消息消费队列在同一时刻只会被消费组内一个消费者消费，一个消费者同一时刻可以分配多个消费队列。

代码清单 5-85　RebalanceImpl#updateProcessQueueTableInRebalance

```
List<PullRequest> pullRequestList = new ArrayList<PullRequest>();
for (MessageQueue mq : mqSet) {
    if (!this.processQueueTable.containsKey(mq)) {
        if (isOrder && !this.lock(mq)) {
            log.warn("doRebalance, {}, add a new mq failed, {}, because lock failed",
                consumerGroup, mq);
            continue;
        }
        this.removeDirtyOffset(mq);
        ProcessQueue pq = new ProcessQueue();
            // 省略部门代码
        long nextOffset = this.computePullFromWhere(mq);
        PullRequest pullRequest = new PullRequest();
        pullRequest.setProcessQueue(pq);
        pullRequestList.add(pullRequest);
    }
}
```

如果经过消息队列重新负载（分配）后，分配到新的消息队列时，首先需要尝试向 Broker 发起锁定该消息队列的请求，如果返回加锁成功则创建该消息队列的拉取任务，否则将跳过，等待其他消费者释放该消息队列的锁，然后在下一次队列重新负载时再尝试加锁。加锁逻辑在下文重点介绍。

顺序消息消费与并发消息消费的第一个关键区别：顺序消息在创建消息队列拉取任务时需要在 Broker 服务器锁定该消息队列。

5.9.2 消息拉取

RocketMQ 消息拉取由 PullMessageService 线程负责，根据消息拉取任务循环拉取消息。

代码清单 5-86　DefaultMQPushConsumerImpl#pullMessage

```
if (this.consumeOrderly) {
    if (processQueue.isLocked()) {
        if (!pullRequest.isLockedFirst()) {
            final long offset = this.rebalanceImpl.computePullFromWhere
                (pullRequest.getMessageQueue());
            boolean brokerBusy = offset < pullRequest.getNextOffset();
            log.info("the first time to pull message, so fix offset from broker.
                pullRequest: {} NewOffset: {} brokerBusy: {}",
```

```
                            pullRequest, offset, brokerBusy);
            if (brokerBusy) {
                log.info("[NOTIFYME]the first time to pull message, but pull request
                    offset larger than broker consume offset. pullRequest: {} NewOffset:
                    {}",pullRequest, offset);
            }
            pullRequest.setLockedFirst(true);
            pullRequest.setNextOffset(offset);
        }
    } else {
        this.executePullRequestLater(pullRequest,
                PULL_TIME_DELAY_MILLS_WHEN_EXCEPTION);
        log.info("pull message later because not locked in broker, {}", pullRequest);
        return;
    }
}
```

如果消息处理队列未被锁定，则延迟 3s 后再将 PullRequest 对象放入到拉取任务中，如果该处理队列是第一次拉取任务，则首先计算拉取偏移量，然后向消息服务端拉取消息。

5.9.3 消息消费

顺序消息消费的实现类：org.apache.rocketmq.client.impl.consumer.ConsumeMessageOrderlyServiCe。其核心类图如图 5-24 所示。

```
ConsumeMessageOrderlyService
-private final DefaultMQPushConsumerImpl defaultMQPushConsumerImpl
-private final DefaultMQPushConsumer defaultMQPushConsumer
-private final MessageListenerOrderly messageListener
-private final BlockingQueue consumeRequestQueue
-private final ThreadPoolExecutor consumeExecutor
-private final String consumerGroup
-private final MessageQueueLock messageQueueLock
-private final ScheduledExecutorService scheduledExecutorService
-private volatile boolean stopped = false
```

图 5-24 ConsumeMessageOrderlyService 类图

下面让我们一一来介绍 ConsumeMessageOrderlyService 的核心属性。

1）MAX_TIME_CONSUME_CONTINUOUSLY：每次消费任务最大持续时间，默认为 60s，可以通过 -Drocketmq.client.maxTimeConsumeContinuously 改变默认值。

2）DefaultMQPushConsumerImpl defaultMQPushConsumerImpl：消息消费者实现类。

3）DefaultMQPushConsumer defaultMQPushConsumer：消息消费者。

4）MessageListenerOrderly messageListener：顺序消息消费监听器。

5）BlockingQueue<Runnable> consumeRequestQueue：消息消费任务队列。

6）ThreadPoolExecutor consumeExecutor：消息消费线程池。

7）String consumerGroup：消息组名。

8）MessageQueueLock messageQueueLock：消息消费端消息消费队列锁容器，内部持有 ConcurrentMap<MessageQueue, Object> mqLockTable =new ConcurrentHashMap<MessageQueue, Object>()。

9）ScheduledExecutorService scheduledExecutorService：调度任务线程池。

1. ConsumeMessageOrderlyService 构造方法

代码清单 5-87　ConsumeMessageOrderlyService 构造方法

```
this.consumeExecutor = new ThreadPoolExecutor(
    this.defaultMQPushConsumer.getConsumeThreadMin(),
    this.defaultMQPushConsumer.getConsumeThreadMax(),1000 * 60,
    TimeUnit.MILLISECONDS,this.consumeRequestQueue,
    new ThreadFactoryImpl("ConsumeMessageThread_"));
this.scheduledExecutorService =
    Executors.newSingleThreadScheduledExecutor(new
        ThreadFactoryImpl("ConsumeMessageScheduledThread_"));
}
```

初始化实例参数，这里的关键是消息任务队列为 LinkedBlockingQueue，消息消费线程池最大运行时线程个数为 consumeThreadMin，consumeThreadMax 参数将失效。

2. ConsumeMessageOrderlyService 启动方法

代码清单 5-88　ConsumeMessageOrderlyService#start

```
public void start() {
    if (MessageModel.CLUSTERING.equals(ConsumeMessageOrderlyService.this.
            defaultMQPushConsumerImpl.messageModel())) {
        this.scheduledExecutorService.scheduleAtFixedRate(new Runnable() {
            public void run() {
                ConsumeMessageOrderlyService.this.lockMQPeriodically();
            }
        }, 1000 * 1, ProcessQueue.REBALANCE_LOCK_INTERVAL, TimeUnit.MILLISECONDS);
    }
}
```

如果消费模式为集群模式，启动定时任务，默认每隔 20s 执行一次锁定分配给自己的消息消费队列。通过 -Drocketmq.client.rebalance.lockInterval=20000 设置间隔，该值建议与一次消息负载频率设置相同。从上文可知，集群模式下顺序消息消费在创建拉取任务时并未将 ProcessQueue 的 locked 状态设置为 true，在未锁定消息队列之前无法执行消息拉取任务，ConsumeMessageOrderlyService 以每 20s 的频率对分配给自己的消息队列进行自动加锁操作，从而消费加锁成功的消息消费队列。接下来分析一下解锁的具体实现。

代码清单 5-89　RebalanceImpl#buildProcessQueueTableByBrokerName

```
private HashMap<String/* brokerName */, Set<MessageQueue>>
    buildProcessQueueTableByBrokerName() {
```

```java
HashMap<String, Set<MessageQueue>> result = new HashMap<String,
    Set<MessageQueue>>();
for (MessageQueue mq : this.processQueueTable.keySet()) {
    Set<MessageQueue> mqs = result.get(mq.getBrokerName());
    if (null == mqs) {
        mqs = new HashSet<MessageQueue>();
        result.put(mq.getBrokerName(), mqs);
    }
    mqs.add(mq);
}
return result;
}
```

Step1：ConcurrentMap<MessageQueue, ProcessQueue> processQueueTable，将消息队列按照 Broker 组织成 Map<String/*brokerName*/,Set<MessageQueue>>，方便下一步向 Broker 发送锁定消息队列请求。

代码清单 5-90　RebalanceImpl#lockAll

```java
LockBatchRequestBody requestBody = new LockBatchRequestBody();
requestBody.setConsumerGroup(this.consumerGroup);
requestBody.setClientId(this.mQClientFactory.getClientId());
requestBody.setMqSet(mqs);
Set<MessageQueue> lockOKMQSet = this.mQClientFactory.
    getMQClientAPIImpl().lockBatchMQ(findBrokerResult.getBrokerAddr(),
        requestBody, 1000);
```

Step2：向 Broker（Master 主节点）发送锁定消息队列，该方法返回成功被当前消费者锁定的消息消费队列。

代码清单 5-91　RebalanceImpl#lockAll

```java
for (MessageQueue mq : lockOKMQSet) {
    ProcessQueue processQueue = this.processQueueTable.get(mq);
    if (processQueue != null) {
    if (!processQueue.isLocked()) {
        log.info("the message queue locked OK, Group: {} {}", this.consumerGroup, mq);
    }
    processQueue.setLocked(true);
    processQueue.setLastLockTimestamp(System.currentTimeMillis());
    }
}
```

Step3：将成功锁定的消息消费队列相对应的处理队列设置为锁定状态，同时更新加锁时间。

代码清单 5-92　RebalanceImpl#lockAll

```java
for (MessageQueue mq : mqs) {
    if (!lockOKMQSet.contains(mq)) {
        ProcessQueue processQueue = this.processQueueTable.get(mq);
```

```
        if (processQueue != null) {
            processQueue.setLocked(false);
            log.warn("the message queue locked Failed, Group: {} {}",
                this.consumerGroup, mq);
        }
    }
}
```

Step4：遍历当前处理队列中的消息消费队列，如果当前消费者不持有该消息队列的锁，将处理队列锁状态设置为 false，暂停该消息消费队列的消息拉取与消息消费。

3. ConsumeMessageOrderlyService 提交消费任务

代码清单 5-93　ConsumeMessageOrderlyService#submitConsumeRequest

```
public void submitConsumeRequest(final List<MessageExt> msgs,final ProcessQueue
    processQueue,final MessageQueue messageQueue,final boolean dispathToConsume)
{
    if (dispathToConsume) {
        ConsumeRequest consumeRequest = new ConsumeRequest(processQueue,
            messageQueue);
        this.consumeExecutor.submit(consumeRequest);
    }
}
```

构建消费任务 ConsumeRequest，并提交到消费线程池中。ConsumeRequest 类图如图 5-25 所示。

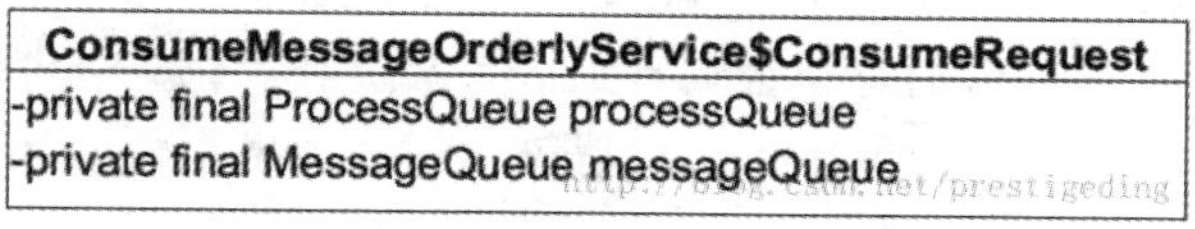

```
ConsumeMessageOrderlyService$ConsumeRequest
-private final ProcessQueue processQueue
-private final MessageQueue messageQueue
```

图 5-25　ConsumeMessageOrderlyService$ConsumeRequest 类图

顺序消息的 ConsumeRequest 消费任务不会直接消费本次拉取的消息，而是在消息消费时从处理队列中拉取消息，接下来详细分析一下 ConsumeRequest 的 run 方法。

代码清单 5-94　ConsumeMessageOrderlyService$ConsumeRequest#run

```
if (this.processQueue.isDropped()) {
    log.warn("run, the message queue not be able to consume, because it's dropped.
        {}", this.messageQueue);
    return;
}
```

Step1：如果消息处理队列为丢弃，则停止本次消费任务。

代码清单 5-95　ConsumeMessageOrderlyService$ConsumeRequest#run

```
final Object objLock = messageQueueLock.fetchLockObject(this.messageQueue);
synchronized (objLock) {
```

Step2：根据消息队列获取一个对象。然后消息消费时先申请独占 objLock。顺序消息消费的并发度为消息队列。也就是一个消息消费队列同一时刻只会被一个消费线程池中一个线程消费。

代码清单 5-96　ConsumeMessageOrderlyService$ConsumeRequest#run

```
if(MessageModel.BROADCASTING.equals(ConsumeMessageOrderlyService.this.default
    MQPushConsumerImpl.messageModel()) ||
    (this.processQueue.isLocked() && !this.processQueue.isLockExpired())) {
    // 消息消费逻辑
} else {
    if (this.processQueue.isDropped()) {
        log.warn("the message queue not be able to consume, because it's dropped.
            {}", this.messageQueue);
        return;
    }
    ConsumeMessageOrderlyService.this.tryLockLaterAndReconsume(
        this.messageQueue, this.processQueue, 100);
}
```

Step3：如果是广播模式的话，直接进入消费，无须锁定处理队列，因为相互直接无竞争；如果是集群模式，消息消费的前提条件是 proceessQueue 被锁定并且锁未超时。思考一下，会不会出现当消息队列重新负载时，原先由自己处理的消息队列被另外一个消费者分配，此时如果还未来得及将 ProceeQueue 解除锁定，就被另外一个消费者添加进去，此时会存储多个消息消费者同时消费一个消息队列？答案是不会的，因为当一个新的消费队列分配给消费者时，在添加其拉取任务之前必须先向 Broker 发送对该消息队列加锁请求，只有加锁成功后，才能添加拉取消息，否则等到下一次负载后，只有消息队列被原先占有的消费者释放后，才能开始新的拉取任务。集群模式下，如果未锁定处理队列，则延迟该队列的消息消费。

代码清单 5-97　ConsumeMessageOrderlyService$ConsumeRequest#run

```
final long beginTime = System.currentTimeMillis();
for (boolean continueConsume = true; continueConsume; ) {
    ... 省略相关代码
    long interval = System.currentTimeMillis() - beginTime;
    if (interval > MAX_TIME_CONSUME_CONTINUOUSLY) {
        ConsumeMessageOrderlyService.this.submitConsumeRequestLater(
            processQueue, messageQueue, 10);
        break;
    }
}
```

Step4：顺序消息消费处理逻辑，每一个 ConsumeRequest 消费任务不是以消费消息条数来计算的，而是根据消费时间，默认当消费时长大于 MAX_TIME_CONSUME_CONTINUOUSLY，默认 60s 后，本次消费任务结束，由消费组内其他线程继续消费。

代码清单 5-98　ConsumeMessageOrderlyService$ConsumeRequest#run

```
final int consumeBatchSize = ConsumeMessageOrderlyService.this.
    defaultMQPushConsumer.getConsumeMessageBatchMaxSize();
List<MessageExt> msgs = this.processQueue.takeMessags(consumeBatchSize);
```

Step5：每次从处理队列中按顺序取出 consumeBatchSize 消息，如果未取到消息，则设置 continueConsume 为 false，本次消费任务结束。顺序消息消费时，从 ProceessQueue 中取出的消息，会临时存储在 ProceeQueue 的 consumingMsgOrderlyTreeMap 属性中。

Step6：执行消息消费钩子函数（消息消费之前 before 方法），通过 DefaultMQPushConsumerImpl#registerConsumeMessageHook（ConsumeMessageHook consumeMessagehook）注册消息消费钩子函数并可以注册多个。

代码清单 5-99　ConsumeMessageOrderlyService$ConsumeRequest#run

```
long beginTimestamp = System.currentTimeMillis();
ConsumeReturnType returnType = ConsumeReturnType.SUCCESS;
boolean hasException = false;
try {
    this.processQueue.getLockConsume().lock();
    if (this.processQueue.isDropped()) {
        log.warn("consumeMessage, the message queue not be able to consume, because
            it's dropped. {}", this.messageQueue);
        break;
    }
    status = messageListener.consumeMessage(
            Collections.unmodifiableList(msgs), context);
} catch (Throwable e) {
    hasException = true;
} finally {
    this.processQueue.getLockConsume().unlock();
}
```

Step7：申请消息消费锁，如果消息队列被丢弃，放弃该消息消费队列的消费，然后执行消息消费监听器，调用业务方具体消息监听器执行真正的消息消费处理逻辑，并通知 RocketMQ 消息消费结果。

Step8：执行消息消费钩子函数，计算消息消费过程中应用程序抛出异常，钩子函数的后处理逻辑也会被调用。

Step9：如果消息消费结果为 ConsumeOrderlyStatus.SUCCESS，执行 ProceeQueue 的 commit 方法，并返回待更新的消息消费进度。

代码清单 5-100　ProcessQueue#commit

```
public long commit() {
    try {
        this.lockTreeMap.writeLock().lockInterruptibly();
        try {
```

```
                Long offset = this.msgTreeMapTemp.lastKey();
                msgCount.addAndGet(this.msgTreeMapTemp.size() * (-1));
                this.msgTreeMapTemp.clear();
                if (offset != null) {
                    return offset + 1;
                }
        } finally {
            this.lockTreeMap.writeLock().unlock();
        }
    } catch (InterruptedException e) {
        log.error("commit exception", e);
    }
    return -1;
}
```

提交,就是将该批消息从 ProceeQueue 中移除,维护 msgCount (消息处理队列中消息条数)并获取消息消费的偏移量 offset,然后将该批消息从 msgTreeMapTemp 中移除,并返回待保存的消息消费进度(offset+1),从中可以看出 offset 表示消息消费队列的逻辑偏移量,类似于数组的下标,代表第 n 个 ConsumeQueue 条目。

代码清单 5-101　ConsumeMessageOrderlyService#processConsumeResult

```
if (checkReconsumeTimes(msgs)){
    consumeRequest.getProcessQueue().makeMessageToCosumeAgain(msgs);
        this.submitConsumeRequestLater(consumeRequest.getProcessQueue(),
            consumeRequest.getMessageQueue(),
            context.getSuspendCurrentQueueTimeMillis());
        continueConsume = false;
} else {
        commitOffset = consumeRequest.getProcessQueue().commit();
}
```

1)检查消息的重试次数。如果消息重试次数大于或等于允许的最大重试次数,将该消息发送到 Broker 端,该消息在消息服务端最终会进入到 DLQ (死信队列),也就是 RocketMQ 不会再次消费,需要人工干预。如果消息成功进入到 DLQ 队列,checkReconsumeTimes 返回 false,该批消息将直接调用 ProcessQueue#commit 提交,表示消息消费成功,如果这批消息中有任意一条消息的重试次数小于允许的最大重试次数,将返回 true,执行消息重试。

> **注意** RocketMQ 顺序消息消费,如果消息重试次数达到允许的最大重试次数并且向 Broker 服务器发送 ACK 消息返回成功,也就是成功将该消息存入到 RocketMQ 的 DLQ 队列中即认为是消息消费成功,继续该消息消费队列后续消息的消费。

2)消息消费重试,先将该批消息重新放入到 ProcessQueue 的 msgTreeMap,然后清除 consumingMsgOrderlyTreeMap,默认延迟 1s 再加入到消费队列中,并结束此次消息消费。

可以通过 DefaultMQPushConsumer#setSuspendCurrentQueueTimeMillis 设置当前队列重试挂起时间。如果执行消息重试，因为消息消费进度并未向前推进，故本次视为无效消费，将不更新消息消费进度。

Step10：存储消息消费进度。

5.9.4 消息队列锁实现

顺序消息消费的各个环节基本都是围绕消息消费队列（MessageQueue）与消息处理队列（ProceeQueue）展开的。消息消费进度拉取，消息进度消费都要判断 ProceeQueue 的 locked 是否为 true，设置 ProceeQueue 为 true 的前提条件是消息消费者（cid）向 Broker 端发送锁定消息队列的请求并返回加锁成功。服务端关于 MessageQueue 加锁处理类：org.apache.rocketmq.broker.client.rebalance.RebalanceLockManager。类图如图 5-26 所示。

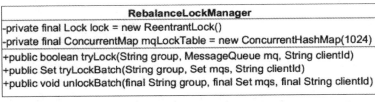

图 5-26　RebalanceLockManager 类图

1）REBALANCE_LOCK_MAX_LIVE_TIME：锁最大存活时间。可以通过 -Drocketmq.broker.rebalance.lockMaxLiveTime 设置，默认为 60s。

2）ConcurrentMap<String/* group */, ConcurrentHashMap<MessageQueue, LockEntry>> mqLockTable：锁容器，以消息消费组分组，每个消息队列对应一个锁对象，表示当前该消息队列被消费组中哪个消费者所持有。

核心方法如下。

1）public Set tryLockBatch（String group, Set<MessageQueue>mqs, String clientId）

申请对 mqs 消息消费队列集合加锁。

String group：消息消费组名。

Set<MessageQueue> mqs：待加锁的消息消费队列集合。

String clientId：消息消费者（cid）。

返回成功加锁的消息队列集合。

2）public void unlockBatch（String group, Set<MessageQueue> mqs, String clientId）

申请对 mqs 消息消费队列集合解锁。

String group：消息消费组。
Set<MessageQueue> mqs：待解锁消息队列集合。
Stirng clientId：持有锁的消息消费者。

由于上述方法都是对 ConcurrentMap<String/* group */, ConcurrentHashMap<MessageQueue, LockEntry>> mqLockTable 数据结构的维护，实现简单，故未对其进行源码分析。

5.10　本章小结

本章主要介绍了消息消费的实现细节。其主要关注点包括消息消费方式、消息队列负载、消息拉取、消息消费、消息消费进度存储、消息过滤、定时消息、顺序消息。

RocketMQ 消息消费方式分别为集群模式与广播模式、集群模式。

消息队列负载由 RebalanceService 线程默认每隔 20s 进行一次消息队列负载，根据当前消费组内消费者个数与主题队列数量按照某一种负载算法进行队列分配，分配原则为同一个消费者可以分配多个消息消费队列，同一个消息消费队列同一时间只会分配给一个消费者。

消息拉取由 PullMessageService 线程根据 RebalanceService 线程创建的拉取任务进行拉取，默认一批拉取 32 条消息，提交给消费者消费线程池后继续下一次的消息拉取。如果消息消费过慢产生消息堆积会触发消息消费拉取流控。PullMessageServicve 与 Rebalance-Service 线程的交互图如图 5-27 所示。

并发消息消费指消费线程池中的线程可以并发地对同一个消息消费队列的消息进行消费，消费成功后，取出消息处理队列中最小的消息偏移量作为消息消费进度偏移量存在于消息消费进度存储文件中，集群模式消息进度存储在 Broker（消息服务器），广播模式消息进度存储在消费者端。如果业务方返回 RECONSUME_LATER，则 RocketMQ 启用消息消费重试机制，将原消息的主题与队列存储在消息属性中，将消息存储在主题名为 SCHEDULE_TOPIC_XXXX 的消息消费队列中，等待指定时间后，RocketMQ 将自动将该消息重新拉取并再次将消息存储在 commitlog 进而转发到原主要的消息消费队列供消费者消费，消息消费重试主题为 %RETRY% 消费者组名。

RocketMQ 不支持任意精度的定时调度消息，只支持自定义的消息延迟级别，例如 1s 2s 5s 等，可通过在 broker 配置文件中设置 messageDelayLevel。其实现原理是 RocketMQ 为这些延迟级别定义对应的消息消费队列，其主题为 SCHEDULE_TOPIC_XXXX，然后创建对应延迟级别的定时任务从消息消费队列中将消息拉取并恢复消息的原主题与原消息消费队列再次存入 commitlog 文件并转发到相应的消息消费队列以便消息消费者拉取消息并消费。

RocketMQ 消息消费支持表达式与类过滤模式，本章重点分析了基于表达式的消息过滤，其中表达式消息过滤又分为基于 TAG 模式与 SQL92 表达式，TAG 模式就是为消息设定一个 TAG，然后消息消费者订阅 TAG，如果消费者订阅的 TAG 列表包含消息的 TAG 则消费该消息。SQL92 表达式基于消息属性实现 SQL 条件表达式的过滤模式。

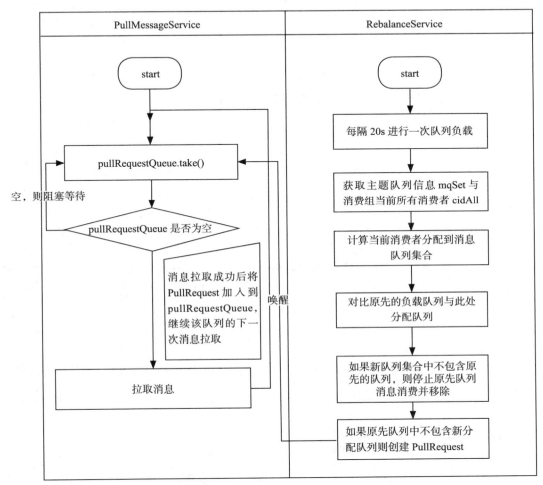

图 5-27　PullMessageService 线程与 RebalanceService 线程交互图

顺序消息消费一般使用集群模式，是指消息消费者内的线程池中的线程对消息消费队列只能串行消费。与并发消息消费最本质的区别是消费消息时必须成功锁定消息消费队列，在 Broker 端会存储消息消费队列的锁占用情况。

Chapter 6 第 6 章

消息过滤 FilterServer

在消息消费的时候,我们会考虑到各种情况,并不是所有消息都需要进行消费的,需要查询出包含特殊标志的消息进行消费,而本章主要分析 RocketMQ 基于类模式的消息过滤机制,主要内容如下。

- ClassFilter 运行机制
- FilterClass 订阅信息注册
- FilterServer 注册剖析
- 消息拉取(拉模式)

6.1 ClassFilter 运行机制

RocketMQ 提供了基于表达式与基于类模式两种过滤模式,在第 5 章已经详细介绍了整个消息拉取、基于表达式(TAG)的过滤模式。基于类模式过滤是指在 Broker 端运行 1 个或多个消息过滤服务器(FilterServer),RocketMQ 允许消息消费者自定义消息过滤实现类并将其代码上传到 FilterServer 上,消息消费者向 FilterServer 拉取消息,FilterServer 将消息消费者的拉取命令转发到 Broker,然后对返回的消息执行消息过滤逻辑,最终将消息返回给消费端,其工作原理如图 6-1 所示。

1)Broker 进程所在的服务器会启动多个 FilterServer 进程。

2)消费者在订阅消息主题时会上传一个自定义的消息过滤实现类,FilterServer 加载并实例化。

3)消息消费者(Consume)向 FilterServer 发送消息拉取请求,FilterServer 接收到消息

消费者消息拉取请求后，FilterServer 将消息拉取请求转发给 Broker，Broker 返回消息后在 FilterServer 端执行消息过滤逻辑，然后返回符合订阅信息的消息给消息消费者进行消费。

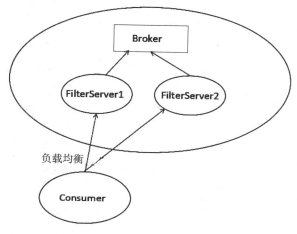

图 6-1　FilterServer 工作原理图

通常消息消费者是直接向 Broker 订阅主题然后从 Broker 上拉取消息，类模式的一个特别之处在于消息消费者是从 FilterServer 拉取消息，那消息消费者是如何感知 FilterServer 的呢？带着该疑问，让我们开始 RocketMQ 类模式消息过滤的学习。

6.2　FilterServer 注册剖析

FilterServer 在启动时会创建一个定时调度任务，每隔 10s 向 Broker 注册自己，请参考代码清单 6-1。

代码清单 6-1　FiltersrvController#initialize

```
this.scheduledExecutorService.scheduleAtFixedRate(new Runnable() {
    public void run() {
        FiltersrvController.this.registerFilterServerToBroker();
    }
}, 3, 10, TimeUnit.SECONDS);
```

Step1：FilterServer 从配置文件中获取 Broker 地址，然后将 FilterServer 所在机器的 IP 与监听端口发送到 Broker 服务器，请求命令类型为 RequestCode.REGISTER_FILTER_SERVER，具体细节如下。

代码清单 6-2　FiltersrvController#registerFilterServerToBroker

```
RegisterFilterServerResponseHeader responseHeader =
    this.filterServerOuterAPI.registerFilterServerToBroker(
        this.filtersrvConfig.getConnectWhichBroker(), this.localAddr());
```

Step2：在 Broker 端处理 REGISTER_FILTER_SERVER 命令的核心实现为 FilterServer-Manager，其实现过程是先从 filterServerTable 中以网络通道为 key 获取 FilterServerInfo，如果不等于空，则更新一下上次更新时间为当前时间，否则创建一个新的 FilterServerInfo 对象并加入到 filterServerTable 路由表中。代码清单 6-3 就是此实现过程的方法。

代码清单 6-3　FilterServerManager#registerFilterServer

```
public void registerFilterServer(final Channel channel, final String
        filterServerAddr) {
    FilterServerInfo filterServerInfo = this.filterServerTable.get(channel);
    if (filterServerInfo != null) {
        filterServerInfo.setLastUpdateTimestamp(System.currentTimeMillis());
    } else {
        filterServerInfo = new FilterServerInfo();
        filterServerInfo.setFilterServerAddr(filterServerAddr);
        filterServerInfo.setLastUpdateTimestamp(System.currentTimeMillis());
        this.filterServerTable.put(channel, filterServerInfo);
        log.info("Receive a New Filter Server<{}>", filterServerAddr);
    }
}
```

FilterServerInfo 类图如图 6-2 所示。

1）String filterServerAddr：filterServer 服务器地址。

2）long lastUpdateTimestamp：filterServer 上次发送心跳包时间。

FilterServerInfo
-private String filterServerAddr
-private long lastUpdateTimestamp

图 6-2　FilterServerInfo 类图

FilterServer 与 Broker 通过心跳维持 FilterServer 在 Broker 端的注册，同样在 Broker 每隔 10s 扫描一下该注册表，如果 30s 内未收到 FilterServer 的注册信息，将关闭 Broker 与 FilterServer 的连接。Broker 为了避免 Broker 端 FilterServer 的异常退出导致 FilterServer 进程越来越少，同样提供一个定时任务每 30s 检测一下当前存活的 FilterServer 进程的个数，如果当前存活的 FilterServer 进程个数小于配置的数量，则自动创建一个 FilterrServer 进程，其具体实现过程如下。

代码清单 6-4　FilterServerManager#createFilterServer

```
public void createFilterServer() {
    int more = this.brokerController.getBrokerConfig().getFilterServerNums() -
            this.filterServerTable.size();
    String cmd = this.buildStartCommand();
    for (int i = 0; i < more; i++) {
        FilterServerUtil.callShell(cmd, log);
    }
}
```

Step1：读取配置文件中的属性 filterServerNums，如果当前运行的 FilterServer 进程数量小于 filterServerNums 则构建 shell 命名并调用。

代码清单 6-5　FilterServerManager#buildStartCommand

```java
private String buildStartCommand() {
    String config = "";
    if (BrokerStartup.configFile != null) {
        config = String.format("-c %s", BrokerStartup.configFile);
    }
    if (this.brokerController.getBrokerConfig().getNamesrvAddr() != null) {
        config += String.format(" -n %s", this.brokerController.getBrokerConfig().getNamesrvAddr());
    }
    if (RemotingUtil.isWindowsPlatform()) {
        return String.format("start /b %s\\bin\\mqfiltersrv.exe %s",
            this.brokerController.getBrokerConfig().getRocketmqHome(),config);
    } else {
        return String.format("sh %s/bin/startfsrv.sh %s",
            this.brokerController.getBrokerConfig().getRocketmqHome(),config);
    }
}
```

Step2：构建启动命令，具体请参考代码清单 6-6。

代码清单 6-6　FilterServerUtil#callShell

```java
public static void callShell(final String shellString, final Logger log) {
    Process process = null;
    try {
        String[] cmdArray = splitShellString(shellString);
        process = Runtime.getRuntime().exec(cmdArray);
        process.waitFor();
        log.info("CallShell: <{}> OK", shellString);
    } catch (Throwable e) {
        log.error("CallShell: readLine IOException, {}", shellString, e);
    } finally {
        if (null != process)
            process.destroy();
    }
}
```

Step3：利用 Runtime.getRuntime() 直接指向 shell 脚本，这里主要是向读者展示一下 JAVA 调用 shell 脚本的一种实现方法。

经过上面的步骤，Broker 上已经保存了 FilterServer 的信息。回想一下第 2 章在探讨 NameServer 存储主题的路由信息中路由元数据中包含 private final HashMap<String/* brokerAddr */, List<String>/* Filter Server */> filterServerTable，那么 NameServer 中关于 Broker 的 filterServer 信息是如何从消息服务器（Broker）传输到 NameServer 的呢？答案是

通过 Broker 与 NameServer 的心跳包来实现。

代码清单 6-7　BrokerOuterAPI#registerBrokerAll

```
if (nameServerAddressList != null) {
    for (String namesrvAddr : nameServerAddressList) {
        try {
            RegisterBrokerResult result = this.registerBroker(namesrvAddr,
                clusterName, brokerAddr, brokerName, brokerId,
                haServerAddr, topicConfigWrapper, filterServerList, oneway,
                timeoutMills);
            if (result != null) {
                registerBrokerResult = result;
            }
            log.info("register broker to name server {} OK", namesrvAddr);
        } catch (Exception e) {
            log.warn("registerBroker Exception, {}", namesrvAddr, e);
        }
    }
}
```

Broker 每 30s 向所有 NameServer 发送心跳包，心跳包中包含了集群名称、Broker 名称、Broker 地址、BrokerId、haServer 地址、topic 配置、过滤服务器列表等。

FilterServer 在启动时向 Broker 注册自己，在 Broker 端维护该 Broker 的 FilterServer 信息，并定时监控 FilterServer 的状态，然后 Broker 通过与所有 NameServer 的心跳包向 NameServer 注册 Broker 上存储的 FilterServer 列表，指引消息消费者正确从 FilterServer 上拉取消息。

6.3　类过滤模式订阅机制

RocketMQ 通过 DefaultMQPushConsumerImpl#subscribe（String topic, String fullClassName, String filterClassSource）方法来实现基于类模式的消息过滤，其参数分别代表消费组订阅的消息主题、类过滤全路径名、类过滤源代码字符串。

代码清单 6-8　DefaultMQPushConsumerImpl#subscribe

```
SubscriptionData subscriptionData = FilterAPI.buildSubscriptionData
    (this.defaultMQPushConsumer.getConsumerGroup(),topic, subExpression);
this.rebalanceImpl.getSubscriptionInner().put(topic, subscriptionData);
```

Step1：构建订阅信息，然后将该订阅信息添加到 RebalanceImpl 中，其主要目标是 RebalanceImpl 会对订阅信息表中的主题进行消息队列的负载，创建消息拉取任务，以便 PullMessageService 线程拉取消息。

代码清单 6-9　MQClientInstance#sendHeartbeatToAllBrokerWithLock
```
this.sendHeartbeatToAllBroker();
this.uploadFilterClassSource();
```

Step2：定时将消息端订阅信息中的类过滤模式的过滤类源码上传到 FilterServer，过程如下。

代码清单 6-10　MQClientInstance#uploadFilterClassToAllFilterServer
```
TopicRouteData topicRouteData = this.topicRouteTable.get(topic);
if (topicRouteData != null && topicRouteData.getFilterServerTable() != null
    && !topicRouteData.getFilterServerTable().isEmpty()) {
// 省略上传相关代码
}
```

Step3：根据订阅的主题获取该主题的路由信息，如果该主题路由信息中的 FilterServer 缓存表不为空，则需要将过滤类发送到 FilterServer 上。TopicRouteData 中 filterServerTable 缓存表的存储格式为 HashMap<String/* brokerAddr */, List<String>/* Filter Server */>，FilterServer 是依附于 Broker 消息服务器的，多个 FilterServer 共同从 Broker 上拉取消息。

代码清单 6-11　MQClientInstance#uploadFilterClassToAllFilterServer
```
for (final String fsAddr : value) {
    try {
        this.mQClientAPIImpl.registerMessageFilterClass(fsAddr, consumerGroup,
            topic, fullClassName, classCRC, classBody,5000);
        log.info("register message class filter to {} OK, ConsumerGroup: {} Topic: {}
            ClassName: {}", fsAddr, consumerGroup,topic, fullClassName);
    } catch (Exception e) {
        log.error("uploadFilterClassToAllFilterServer Exception", e);
    }
}
```

Step4：遍历主题路由表中的 filterServerTable，向缓存中所有的 FilterServer 上传消息过滤代码。

代码清单 6-12　FilterClassManager#registerFilterClass
```
public boolean registerFilterClass(String consumerGroup,String topic,String
    className, int classCRC, byte[] filterSourceBinary)
```

Step5：FilterServer 端处理 FilterClass 上传并将其源码编译的实现为 FilterClass-Manager，该方法的参数含义分别是消息消费组名、消息主题、消息过滤类全路径名、源码的 CRC 验证码、过滤类源码。

代码清单 6-13　FilterClassManager#registerFilterClass
```
final String key = buildKey(consumerGroup, topic);
boolean registerNew = false;
```

```
    FilterClassInfo filterClassInfoPrev = this.filterClassTable.get(key);
    if (null == filterClassInfoPrev) {
        registerNew = true;
    } else if(this.filtersrvController.getFiltersrvConfig().
            isClientUploadFilterClassEnable()) {
        if (filterClassInfoPrev.getClassCRC() != classCRC && classCRC != 0) {
            registerNew = true;
        }
    }
}
```

Step6：根据消息消费组与主题名称构建 filterClasTable 缓存 key，从缓存表中尝试获取过滤类型信息 FilterClassInfo。如果缓存表中不包含 FilterClassInfo 则表示第一次注册，设置 registerNew 为 true；如果 FlterClassInfo 不为空，说明该消息消费组不是第一次注册。如果服务端开启允许消息消费者上传 FilterClass，比较两个的 classCRC，如果不相同，说明 FilterClass 的源码发生了变化，设置 registerNew 为 true。

代码清单 6-14　FilterClassManager#registerFilterClass

```
    FilterClassInfo filterClassInfoNew = new FilterClassInfo();
    filterClassInfoNew.setClassName(className);
    filterClassInfoNew.setClassCRC(0);
    filterClassInfoNew.setMessageFilter(null);
    if (this.filtersrvController.getFiltersrvConfig().
            isClientUploadFilterClassEnable()) {
        String javaSource = new String(filterSourceBinary, MixAll.DEFAULT_CHARSET);
        Class<?> newClass = DynaCode.compileAndLoadClass(className, javaSource);
        Object newInstance = newClass.newInstance();
        filterClassInfoNew.setMessageFilter((MessageFilter) newInstance);
        filterClassInfoNew.setClassCRC(classCRC);
    }
    this.filterClassTable.put(key, filterClassInfoNew);
```

Step7：如果是第一次注册，则创建 FilterClassInfo，如果 FilterServer 允许消息消费者上传过滤类源码，则使用 JDK 提供的方法将源代码编译并加装，然后创建其实例，并强制类型转换为 MessageFilter，也就是自定义的消息过滤类必须实现 MessageFilter 接口。

上述整个过程就完成了消息消费端向 FilterServer 上传过滤类的过程，但如果 FilterServer 不允许消息消费者上传 FilterClass，则 filterServerTable 中存在的过滤类信息只包含 className，classCRC、消息过滤类 MessageFilter 属性都为空，也就是说会忽略消息消费者上传的过滤类源代码，那过滤类的源码从哪获取呢？原来 FilterServer 会开启一个定时任务从配置好的远程服务器去获取过滤类的源码，再将其编译与实例化，其具体实现如下。

代码清单 6-15　FilterClassManager#start

```
    public void start() {
        if (!this.filtersrvController.getFiltersrvConfig().
                isClientUploadFilterClassEnable()) {
```

```
            this.scheduledExecutorService.scheduleAtFixedRate(new Runnable() {
                public void run() {
                    fetchClassFromRemoteHost();
                }
            }, 1, 1, TimeUnit.MINUTES);
        }
    }
```

Step1：如果 FilterServer 不允许消息消费者上传类，也就是不允许直接编译来自消费者上传的类，FilterServer 会开启一个定时任务，每隔 1 分钟从远程服务器下载源代码并编译。

代码清单 6-16　HttpFilterClassFetchMethod#fetch

```
public String fetch(String topic, String consumerGroup, String className) {
    String thisUrl = String.format("%s/%s.java", this.url, className);
    try {
        HttpResult result = HttpTinyClient.httpGet(thisUrl, null, null, "UTF-8",
            5000);
        if (200 == result.code) {
            return result.content;
        }
    } catch (Exception e) {
        // 省略 error 日志输出
    }
    return null;
}
```

Step2：从远程服务器根据消息主题、消息消费组名称、过滤类名称从远程服务器获取。远程服务器的 URL 地址通过 filterClassRepertoryUrl 配置属性指定，该服务器要能响应如下请求连接并返回过滤类的源代码：http://filterClassRepertoryUrl/className.java。

代码清单 6-17　HttpFilterClassFetchMethod#fetchClassFromRemoteHost

```
byte[] filterSourceBinary = responseStr.getBytes("UTF-8");
int classCRC = UtilAll.crc32(responseStr.getBytes("UTF-8"));
if (classCRC != filterClassInfo.getClassCRC()) {
    String javaSource = new String(filterSourceBinary, MixAll.DEFAULT_CHARSET);
    Class<?> newClass = DynaCode.compileAndLoadClass(
        filterClassInfo.getClassName(), javaSource);
    Object newInstance = newClass.newInstance();
    filterClassInfo.setMessageFilter((MessageFilter) newInstance);
    filterClassInfo.setClassCRC(classCRC);
}
```

Step3：获取到过滤类源码后将其编译并创建 MessageFilter 实例且更新 FilterClassInfo。

6.4　消息拉取

RocketMQ 消息的过滤发生在消息消费的时候，PullMessageService 线程默认从 Broker

上拉取消息，执行相关的过滤逻辑，在 FilterServer 过滤模式下，PullMessageService 线程是如何将拉取地址由原来的 Broker 地址转换成 FilterServer 地址呢？

代码清单 6-18　PullAPIWrapper#pullKernelImpl

```
if (PullSysFlag.hasClassFilterFlag(sysFlagInner)) {
    brokerAddr = computPullFromWhichFilterServer(mq.getTopic(), brokerAddr);
}
```

Step1：在消息拉取时，如果发现消息过滤模式为 classFilter，将拉取消息服务器地址由原来的 Broker 地址转换成该 Broker 服务器所对应的 FilterServer。

代码清单 6-19　PullAPIWrapper#computPullFromWhichFilterServer

```
ConcurrentMap<String, TopicRouteData> topicRouteTable =
    this.mQClientFactory.getTopicRouteTable();
if (topicRouteTable != null) {
    TopicRouteData topicRouteData = topicRouteTable.get(topic);
    List<String> list = topicRouteData.getFilterServerTable().get(brokerAddr);
    if (list != null && !list.isEmpty()) {
        return list.get(randomNum() % list.size());
    }
}
```

Step2：获取该消息主题的路由信息，从路由信息中获取 Broker 对应的 FilterServer 列表，如果不为空则随机从 FilterServer 列表中选择一个 FilterServer，发送拉取消息请求至相应的 FilterServer 上，由于 FilterServer 使用 DefaultMQPullConsumer 消费者根据消息消费者的拉取任务将拉取请求转发给 Broker，然后对返回的消息执行消息过滤逻辑，将匹配的消息返回给消息消费者。DefaultMQPullConsumer 拉取机制在第 5 章已详细介绍，在这里就不再重复介绍了。

6.5　本章小结

本章详细介绍了 RocketMQ 另一种消息过滤模式：允许消息消费者在订阅主题消息时上传消息过滤类到过滤服务器，在过滤服务器将消息过滤后再返回给消息消费者，其相比基于 TAG 模式进行消息过滤有如下优势。

1）基于 TAG 模式消息过滤，由于在消息服务端进行消息过滤是匹配消息 TAG 的 hashcode，导致服务端过滤并不十分准确，从服务端返回的消息最终并不一定是消息消费者订阅的消息，造成网络带宽的浪费，而基于类模式的消息过滤所有的过滤操作全部在 FilterServer 端进行。

2）由于 FilterServer 与 Broker 运行在同一台机器上，消息的传输是通过本地回环通信，不会浪费 Broker 端的网络资源。

第 7 章　RocketMQ 主从同步 (HA) 机制

高可用特性是目前分布式系统中必备的特性之一，对一个中间件来说没有 HA 机制是一个重大的缺陷，本章将主要分析 RocketMQ 主从同步（HA）机制。

本章重点内容如下。
- 主从同步复制实现原理。
- RocketMQ 读写分离机制。

7.1　RocketMQ 主从复制原理

为了提高消息消费的高可用性，避免 Broker 发生单点故障引起存储在 Broker 上的消息无法及时消费，RocketMQ 引入了 Broker 主备机制，即消息消费到达主服务器后需要将消息同步到消息从服务器，如果主服务器 Broker 宕机后，消息消费者可以从从服务器拉取消息。

接下来将详细探讨 RocketMQ HA 的实现原理，RocketMQ HA 核心实现类图如图 7-1 所示。

从图 7-1 中我们知道 RocketMQ HA 由 7 个核心类实现，分别如下。

1）HAService：RocketMQ 主从同步核心实现类。

2）HAService$AcceptSocketService：HA Master 端监听客户端连接实现类。

3）HAService$GroupTransferService：主从同步通知实现类。

4）HAService$HAClient：HA Client 端实现类。

5）HAConnection：HA Master 服务端 HA 连接对象的封装，与 Broker 从服务器的网络

读写实现类。

6）HAConnection$ReadSocketService：HA Master 网络读实现类。

7）HAConnection$WriteSocketServicce：HA Master 网络写实现类。

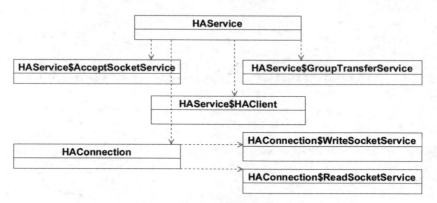

图 7-1　RocketMQ HA 核心类图

7.1.1　HAService 整体工作机制

从 HAService 开始，我们来了解 RocketMQ HA 的工作机制，首先看一下代码清单 7-1。

代码清单 7-1　HAService#start

```
public void start() throws Exception {
    this.acceptSocketService.beginAccept();
    this.acceptSocketService.start();
    this.groupTransferService.start();
    this.haClient.start();
}
```

RocketMQ HA 的实现原理如下。

1）主服务器启动，并在特定端口上监听从服务器的连接。

2）从服务器主动连接主服务器，主服务器接收客户端的连接，并建立相关 TCP 连接。

3）从服务器主动向主服务器发送待拉取消息偏移量，主服务器解析请求并返回消息给从服务器。

4）从服务器保存消息并继续发送新的消息同步请求。

7.1.2　AcceptSocketService 实现原理

AcceptSocketService 作为 HAService 的内部类，实现 Master 端监听 Slave 连接，类图如图 7-2 所示。

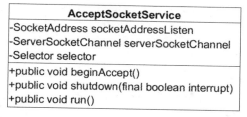

图 7-2 AcceptSocketService 类图

AcceptSocketService 的属性如下。

1）SocketAddress socketAddressListen：Broker 服务监听套接字（本地 IP+ 端口号）。

2）ServerSocketChannel serverSocketChannel：服务端 Socket 通道，基于 NIO。

3）Selector selector：事件选择器，基于 NIO。

代码清单 7-2　HAService$AcceptSocketService#beginAccept

```
public void beginAccept() throws Exception {
    this.serverSocketChannel = ServerSocketChannel.open();
    this.selector = RemotingUtil.openSelector();
    this.serverSocketChannel.socket().setReuseAddress(true);
    this.serverSocketChannel.socket().bind(this.socketAddressListen);
    this.serverSocketChannel.configureBlocking(false);
    this.serverSocketChannel.register(this.selector, SelectionKey.OP_ACCEPT);
}
```

创建 ServerSocketChannel、创建 Selector、设置 TCP reuseAddress、绑定监听端口、设置为非阻塞模式，并注册 OP_ACCEPT（连接事件）。

代码清单 7-3　HAService$AcceptSocketService#run

```
this.selector.select(1000);
    Set<SelectionKey> selected = this.selector.selectedKeys();
    if (selected != null) {
        for (SelectionKey k : selected) {
            if ((k.readyOps() & SelectionKey.OP_ACCEPT) != 0) {
                SocketChannel sc = ((ServerSocketChannel) k.channel()).accept();
                if (sc != null) {
                    HAService.log.info("HAService receive new connection, "+
                        sc.socket().getRemoteSocketAddress());
                    try {
                        HAConnection conn = new HAConnection(HAService.this, sc);
                        conn.start();
                        HAService.this.addConnection(conn);
                    } catch (Exception e) {
                        log.error("new HAConnection exception", e);
                        sc.close();
                    }
                }
            }
        }
    } else {
```

```
                log.warn("Unexpected ops in select " + k.readyOps());
            }
        }
        selected.clear();
    }
```

该方法是标准的基于 NIO 的服务端程式实例，选择器每 1s 处理一次连接就绪事件。连接事件就绪后，调用 ServerSocketChannel 的 accept() 方法创建 SocketChannel。然后为每一个连接创建一个 HAConnection 对象，该 HAConnection 将负责 M-S 数据同步逻辑。

7.1.3　GroupTransferService 实现原理

GroupTransferService 主从同步阻塞实现，如果是同步主从模式，消息发送者将消息刷写到磁盘后，需要继续等待新数据被传输到从服务器，从服务器数据的复制是在另外一个线程 HAConnection 中去拉取，所以消息发送者在这里需要等待数据传输的结果，GroupTransferService 就是实现该功能，该类的整体结构与同步刷盘实现类（CommitLog$GroupCommitService）类似，本节只关注该类的核心业务逻辑 doWaitTransfer 的实现。

代码清单 7-4　HAService$GroupTransferService#doWaitTransfer

```
private void doWaitTransfer() {
    synchronized (this.requestsRead) {
        if (!this.requestsRead.isEmpty()) {
            for (CommitLog.GroupCommitRequest req : this.requestsRead) {
                boolean transferOK = HAService.this.push2SlaveMaxOffset.get() >=
                    req.getNextOffset();
                for (int i = 0; !transferOK && i < 5; i++) {
                    this.notifyTransferObject.waitForRunning(1000);
                    transferOK = HAService.this.push2SlaveMaxOffset.get() >=
                        req.getNextOffset();
                }
                if (!transferOK) {
                 log.warn("transfer messsage to slave timeout, " + req.getNextOffset());
                }
                req.wakeupCustomer(transferOK);
            }
            this.requestsRead.clear();
        }
    }
}
```

GroupTransferService 的职责是负责当主从同步复制结束后通知由于等待 HA 同步结果而阻塞的消息发送者线程。判断主从同步是否完成的依据是 Slave 中已成功复制的最大偏移量是否大于等于消息生产者发送消息后消息服务端返回下一条消息的起始偏移量，如果是则表示主从同步复制已经完成，唤醒消息发送线程，否则等待 1s 再次判断，每一个任务在一批任务中循环判断 5 次。消息发送者返回有两种情况：等待超过 5s 或 GroupTransferService

通知主从复制完成。可以通过 syncFlushTimeout 来设置发送线程等待超时时间。GroupTransferService 通知主从复制的实现如下。

代码清单 7-5 HAService$GroupTransferService#notifyTransferSome

```
public void notifyTransferSome(final long offset) {
    for (long value = this.push2SlaveMaxOffset.get(); offset > value; ) {
        boolean ok = this.push2SlaveMaxOffset.compareAndSet(value, offset);
        if (ok) {
            this.groupTransferService.notifyTransferSome();
            break;
        } else {
            value = this.push2SlaveMaxOffset.get();
        }
    }
}
```

该方法在 Master 收到从服务器的拉取请求后被调用，表示从服务器当前已同步的偏移量，既然收到从服务器的反馈信息，需要唤醒某些消息发送者线程。如果从服务器收到的确认偏移量大于 push2SlaveMaxOffset，则更新 push2SlaveMaxOffset，然后唤醒 GroupTransferService 线程，各消息发送者线程再次判断自己本次发送的消息是否已经成功复制到从服务器。

7.1.4 HAClient 实现原理

HAClient 是主从同步 Slave 端的核心实现类，类图如图 7-3 所示。

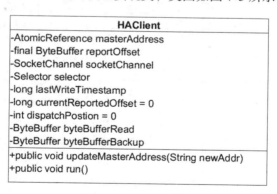

图 7-3 HAClient 类图

HAClient 类的基本属性和常量如下。

1) private static final int READ_MAX_BUFFER_SIZE = 1024 * 1024 * 4：Socket 读缓存区大小。

2) AtomicReference<String> masterAddress：master 地址。

3) ByteBuffer reportOffset = ByteBuffer.allocate（8）：Slave 向 Master 发起主从同步的

拉取偏移量。

4）SocketChannel socketChannel：网络传输通道。
5）Selector selector：NIO 事件选择器。
6）long lastWriteTimestamp：上一次写入时间戳。
7）long currentReportedOffset：反馈 Slave 当前的复制进度，commitlog 文件最大偏移量。
8）dispatchPostion：本次已处理读缓存区的指针。
9）ByteBuffer byteBufferRead：读缓存区，大小为 4M。
10）ByteBuffer byteBufferBackup：读缓存区备份，与 BufferRead 进行交换。

接下来从 run 方法开始探讨 HAClient 的工作原理。

代码清单 7-6　HAClient#connectMaster

```java
private boolean connectMaster() throws ClosedChannelException {
    if (null == socketChannel) {
        String addr = this.masterAddress.get();
        if (addr != null) {
            SocketAddress socketAddress = RemotingUtil.string2SocketAddress(addr);
            if (socketAddress != null) {
                this.socketChannel = RemotingUtil.connect(socketAddress);
                if (this.socketChannel != null) {
                    this.socketChannel.register(this.selector, SelectionKey.OP_READ);
                }
            }
        }
        this.currentReportedOffset =
            HAService.this.defaultMessageStore.getMaxPhyOffset();
        this.lastWriteTimestamp = System.currentTimeMillis();
    }
    return this.socketChannel != null;
}
```

Step1：Slave 服务器连接 Master 服务器。如果 socketChannel 为空，则尝试连接 Master。如果 master 地址为空，返回 false；如果 master 地址不为空，则建立到 Master 的 TCP 连接，然后注册 OP_READ（网络读事件），初始化 currentReportedOffset 为 commitlog 文件的最大偏移量、lastWriteTimestamp 上次写入时间戳为当前时间戳，并返回 true。在 Broker 启动时，如果 Broker 角色为 SLAVE 时将读取 Broker 配置文件中的 haMasterAddress 属性并更新 HAClient 的 masterAddrees，如果角色为 SLAVE 并且 haMasterAddress 为空，启动并不会报错，但不会执行主从同步复制，该方法最终返回是否成功连接上 Master。

代码清单 7-7　HAClient#isTimeToReportOffset

```java
private boolean isTimeToReportOffset() {
    long interval = HAService.this.defaultMessageStore.getSystemClock().now() -
        this.lastWriteTimestamp;
    boolean needHeart = interval > HAService.this.defaultMessageStore.
```

```
            getMessageStoreConfig().getHaSendHeartbeatInterval();
    return needHeart;
}
```

Step2：判断是否需要向 Master 反馈当前待拉取偏移量，Master 与 Slave 的 HA 心跳发送间隔默认为 5s，可通过配置 haSendHeartbeatInterval 来改变心跳间隔。

代码清单 7-8　HAClient#reportSlaveMaxOffset

```
private boolean reportSlaveMaxOffset(final long maxOffset) {
    this.reportOffset.position(0);
    this.reportOffset.limit(8);
    this.reportOffset.putLong(maxOffset);
    this.reportOffset.position(0);
    this.reportOffset.limit(8);
    for (int i = 0; i < 3 && this.reportOffset.hasRemaining(); i++) {
        try {
            this.socketChannel.write(this.reportOffset);
        } catch (IOException e) {
            log.error(this.getServiceName()
                + "reportSlaveMaxOffset this.socketChannel.write exception", e);
            return false;
        }
    }
    return !this.reportOffset.hasRemaining();
}
```

Step3：向 Master 服务器反馈拉取偏移量。这里有两重意义，对于 Slave 端来说，是发送下次待拉取消息偏移量，而对于 Master 服务端来说，既可以认为是 Slave 本次请求拉取的消息偏移量，也可以理解为 Slave 的消息同步 ACK 确认消息。

这里 RocketMQ 作者提供了一个基于 NIO 的网络写示例程序：首先先将 ByteBuffer 的 position 设置为 0，limit 设置为待写入字节长度，然后调用 putLong 将待拉取偏移量写入 ByteBuffer 中，需要将 ByteBuffer 从写模式切换到读模式，这里的用法是手动将 position 设置为 0，limit 设置为可读长度，其实这里可以直接调用 ByteBuffer 的 flip() 方法来切换 ByteBuffer 的读写状态。特别需要留意的是，调用网络通道的 write 方法是在一个 while 循环中反复判断 byteBuffer 是否全部写入到通道中，这是由于 NIO 是一个非阻塞 IO，调用一次 write 方法不一定会将 ByteBuffer 可读字节全部写入。

代码清单 7-9　HAClient#run

```
this.selector.select(1000);
```

Step4：进行事件选择，其执行间隔为 1s。

代码清单 7-10　HAClient#processReadEvent

```java
private boolean processReadEvent() {
    int readSizeZeroTimes = 0;
    while (this.byteBufferRead.hasRemaining()) {
        try {
            int readSize = this.socketChannel.read(this.byteBufferRead);
            if (readSize > 0) {
                lastWriteTimestamp = HAService.this.defaultMessageStore.
                    getSystemClock().now();
                readSizeZeroTimes = 0;
                boolean result = this.dispatchReadRequest();
                if (!result) {
                    log.error("HAClient, dispatchReadRequest error");
                    return false;
                }
            } else if (readSize == 0) {
                if (++readSizeZeroTimes >= 3) {
                    break;
                }
            } else {
                log.info("HAClient, processReadEvent read socket < 0");
                return false;
            }
        } catch (IOException e) {
            log.info("HAClient, processReadEvent read socket exception", e);
            return false;
        }
    }
    return true;
}
```

Step5：处理网络读请求，即处理从 Master 服务器传回的消息数据。同样 RocketMQ 作者给出了一个处理网络读的 NIO 示例。循环判断 readByteBuffer 是否还有剩余空间，如果存在剩余空间，则调用 SocketChannel#read（ByteBuffer readByteBuffer），将通道中的数据读入到读缓存区中。

1）如果读取到的字节数大于 0，重置读取到 0 字节的次数，并更新最后一次写入时间戳（lastWriteTimestamp），然后调用 dispatchReadRequest 方法将读取到的所有消息全部追加到消息内存映射文件中，然后再次反馈拉取进度给服务器。

2）如果连续 3 次从网络通道读取到 0 个字节，则结束本次读，返回 true。

3）如果读取到的字节数小于 0 或发生 IO 异常，则返回 false。

HAClient 线程反复执行上述 5 个步骤完成主从同步复制功能。

7.1.5　HAConnection 实现原理

Master 服务器在收到从服务器的连接请求后，会将主从服务器的连接 SocketChannel 封装成 HAConnection 对象，实现主服务器与从服务器的读写操作。其类图如图 7-4 所示。

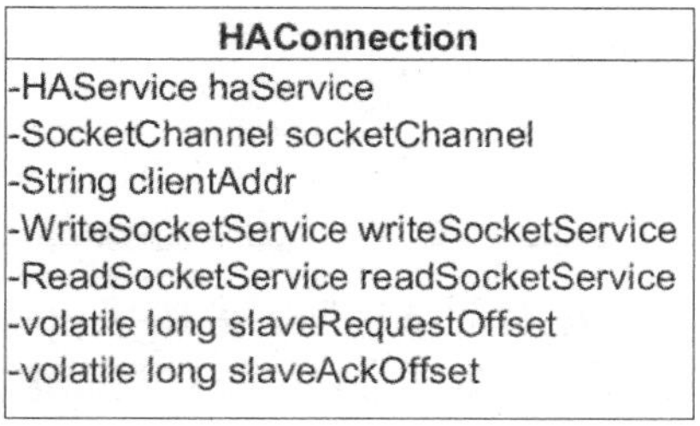

```
HAConnection
-HAService haService
-SocketChannel socketChannel
-String clientAddr
-WriteSocketService writeSocketService
-ReadSocketService readSocketService
-volatile long slaveRequestOffset
-volatile long slaveAckOffset
```

图 7-4　HAConnection 类图

HAConnection 类属性如下。

1）HAService haService：HAService 对象。

2）SocketChannel socketChannel：网络 socket 通道。

3）String clientAddr：客户端连接地址。

4）WriteSocketService writeSocketService：服务端向从服务器写数据服务类。

5）ReadSocketService readSocketService：服务端从从服务器读数据服务类。

6）long slaveRequestOffset：从服务器请求拉取数据的偏移量。

7）long slaveAckOffset：从服务器反馈已拉取完成的数据偏移量。

HAConnection 的网络读请求由其内部类 ReadSocketService 线程来实现，其类图如图 7-5 所示。

```
HAConnection$ReadSocketService
-private static final int READ_MAX_BUFFER_SIZE = 1024 * 1024
-private final Selector selector
-private final SocketChannel socketChannel
-private final ByteBuffer byteBufferRead
-private int processPostion = 0
-private volatile long lastReadTimestamp
```

图 7-5　HAConnection$ReadSocketService 类图

1）READ_MAX_BUFFER_SIZE：网络读缓存区大小，默认为 1M。

2）Selector selector：NIO 网络事件选择器。

3）SocketChannel socketChannel：网络通道，用于读写的 socket 通道。

4）ByteBuffer byteBufferRead：网络读写缓存区，默认为 1M。

5）int processPosition：byteBuffer 当前处理指针。

6）volatile long lastReadTimestamp：上次读取数据的时间戳。

通过观察其 run 方法，每隔 1s 处理一次读就绪事件，每次读请求调用其 processRead-Event 来解析从服务器的拉取请求。接下来将详细剖析其处理读请求的实现细节。

代码清单 7-11　HAConnection$ReadSocketService#processReadEvent

```
int readSizeZeroTimes = 0;
if (!this.byteBufferRead.hasRemaining()) {
    this.byteBufferRead.flip();
    this.processPostion = 0;
}
```

Step1：如果 byteBufferRead 没有剩余空间，说明该 position==limit==capacity，调用 byteBufferRead.flip() 方法，产生的效果为 position=0,limit=capacity 并设置 processPostion 为 0，表示从头开始处理，其实这里调用 byteBuffer.clear() 方法会更加容易理解。

代码清单 7-12　HAConnection$ReadSocketService#processReadEvent

```
while (this.byteBufferRead.hasRemaining()) {
    // 处理网络读
    int readSize = this.socketChannel.read(this.byteBufferRead);
}
```

Step2：NIO 网络读的常规方法，一般使用循环的方式进行读写，直到 byteBuffer 中没有剩余的空间。

代码清单 7-13　HAConnection$ReadSocketService#processReadEvent

```
if (readSize > 0) {
    readSizeZeroTimes = 0;
    this.lastReadTimestamp = HAConnection.this.haService.
        getDefaultMessageStore().getSystemClock().now();
    if ((this.byteBufferRead.position() - this.processPostion) >= 8) {
        int pos = this.byteBufferRead.position() - (this.byteBufferRead.position()
            % 8);
        long readOffset = this.byteBufferRead.getLong(pos - 8);
        this.processPostion = pos;
        HAConnection.this.slaveAckOffset = readOffset;
        if (HAConnection.this.slaveRequestOffset < 0) {
            HAConnection.this.slaveRequestOffset = readOffset;
            log.info("slave[" + HAConnection.this.clientAddr + "] request offset " +
                readOffset);
        }
        HAConnection.this.haService.notifyTransferSome(
            HAConnection.this.slaveAckOffset);
    }
}
```

Step3：如果读取的字节大于 0 并且本次读取到的内容大于等于 8，表明收到了从服务器一条拉取消息的请求。由于有新的从服务器反馈拉取偏移量，服务端会通知由于同步等待 HA 复制结果而阻塞的消息发送者线程。

代码清单 7-14　HAConnection$ReadSocketService#processReadEvent
```
if (readSize == 0) {
    if (++readSizeZeroTimes >= 3) {
        break;
    }
} else {
    log.error("read socket[" + HAConnection.this.clientAddr + "] < 0");
    return false;
}
```

Step4：如果读取到的字节数等于 0，则重复三次，否则结束本次读请求处理；如果读取到的字节数小于 0，表示连接处于半关闭状态，返回 false 则意味着消息服务器将关闭该链接。

HAConnction 的读请求就介绍到这里，其网络写请求由内部类 WriteSocketService 线程来实现，类图如图 7-6 所示。

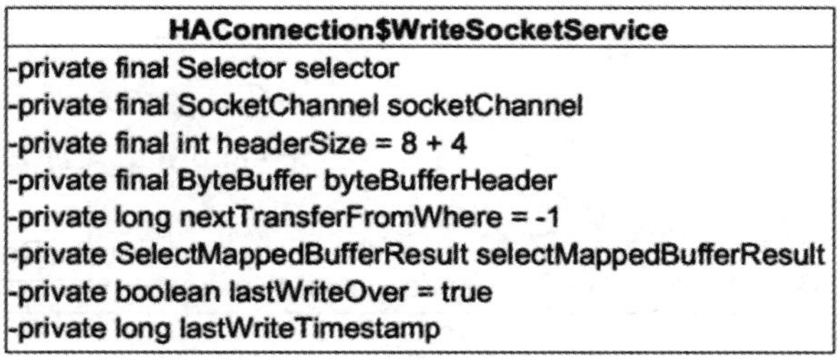

```
HAConnection$WriteSocketService
-private final Selector selector
-private final SocketChannel socketChannel
-private final int headerSize = 8 + 4
-private final ByteBuffer byteBufferHeader
-private long nextTransferFromWhere = -1
-private SelectMappedBufferResult selectMappedBufferResult
-private boolean lastWriteOver = true
-private long lastWriteTimestamp
```

图 7-6　HAConnection$WriteSocketService 类图

1）Selector selector：NIO 网络事件选择器。
2）SocketChannel socketChannel：网络 socket 通道。
3）int headerSize：消息头长度，消息物理偏移量 + 消息长度。
4）long nextTransferFromWhere：下一次传输的物理偏移量。
5）SelectMappedBufferResult selectMappedBufferResult：根据偏移量查找消息的结果。
6）boolean lastWriteOver：上一次数据是否传输完毕。
7）long lastWriteTimestamp：上次写入的时间戳。

接下来分析一下其实现原理，重点关注去 run 方法。

代码清单 7-15　HAConnection$WriteSocketService#run
```
if (-1 == HAConnection.this.slaveRequestOffset) {
    Thread.sleep(10);
    continue;
}
```

Step1：如果 slaveRequestOffset 等于 -1，说明 Master 还未收到从服务器的拉取请求，放弃本次事件处理。slaveRequestOffset 在收到从服务器拉取请求时更新。

代码清单 7-16　HAConnection$WriteSocketService#run

```
if (-1 == this.nextTransferFromWhere) {
    if (0 == HAConnection.this.slaveRequestOffset) {
        long masterOffset = HAConnection.this.haService.getDefaultMessageStore().
                getCommitLog().getMaxOffset();
        masterOffset = masterOffset - (masterOffset %
        HAConnection.this.haService.getDefaultMessageStore().
                getMessageStoreConfig().getMapedFileSizeCommitLog());
        if (masterOffset < 0) {
            masterOffset = 0;
        }
        this.nextTransferFromWhere = masterOffset;
    } else {
        this.nextTransferFromWhere = HAConnection.this.slaveRequestOffset;
    }
}
```

Step2：如果 nextTransferFromWhere 为 -1，表示初次进行数据传输，计算待传输的物理偏移量，如果 slaveRequestOffset 为 0，则从当前 commitlog 文件最大偏移量开始传输，否则根据从服务器的拉取请求偏移量开始传输。

代码清单 7-17　HAConnection$WriteSocketService#run

```
if (this.lastWriteOver) {
    long interval =HAConnection.this.haService.getDefaultMessageStore().
            getSystemClock().now() - this.lastWriteTimestamp;
    if (interval > HAConnection.this.haService.getDefaultMessageStore().
            getMessageStoreConfig().getHaSendHeartbeatInterval()) {
        this.byteBufferHeader.position(0);
        this.byteBufferHeader.limit(headerSize);
        this.byteBufferHeader.putLong(this.nextTransferFromWhere);
        this.byteBufferHeader.putInt(0);
        this.byteBufferHeader.flip();
        this.lastWriteOver = this.transferData();
        if (!this.lastWriteOver)
            continue;
    }
} else {
    this.lastWriteOver = this.transferData();
    if (!this.lastWriteOver)
        continue;
}
```

Step3：判断上次写事件是否已将信息全部写入客户端。

1）如果已全部写入，且当前系统时间与上次最后写入的时间间隔大于 HA 心跳检测时间，则发送一个心跳包，心跳包的长度为 12 个字节（从服务器待拉取偏移量 +size），消息长度默认为 0，避免长连接由于空闲被关闭。HA 心跳包发送间隔通过 haSendHeartbeatInterval 放置，默认值为 5s。

2）如果上次数据未写完，则先传输上一次的数据，如果消息还是未全部传输，则结束此次事件处理。

代码清单 7-18　HAConnection$WriteSocketService#run

```
SelectMappedBufferResult selectResult =
        HAConnection.this.haService.getDefaultMessageStore().getCommitLogData(
                this.nextTransferFromWhere);
if (selectResult != null) {
    int size = selectResult.getSize();
    if (size > HAConnection.this.haService.getDefaultMessageStore().
            getMessageStoreConfig().getHaTransferBatchSize()) {
        size = HAConnection.this.haService.getDefaultMessageStore().
                getMessageStoreConfig().getHaTransferBatchSize();
    }
    long thisOffset = this.nextTransferFromWhere;
    this.nextTransferFromWhere += size;
    selectResult.getByteBuffer().limit(size);
    this.selectMappedBufferResult = selectResult;
    // Build Header
    this.byteBufferHeader.position(0);
    this.byteBufferHeader.limit(headerSize);
    this.byteBufferHeader.putLong(thisOffset);
    this.byteBufferHeader.putInt(size);
    this.byteBufferHeader.flip();
    this.lastWriteOver = this.transferData();
} else {
    HAConnection.this.haService.getWaitNotifyObject().allWaitForRunning(100);
}
```

Step4：传输消息到从服务器。

1）根据消息从服务器请求的待拉取偏移量，查找该偏移量之后所有的可读消息，如果未查到匹配的消息，通知所有等待线程继续等待 100ms。

2）如果匹配到消息，且查找到的消息总长度大于配置 HA 传输一次同步任务最大传输的字节数，则通过设置 ByteBuffer 的 limit 来控制只传输指定长度的字节，这就意味着 HA 客户端收到的消息会包含不完整的消息。HA 一批次传输消息最大字节通过 haTransferBatchSize 设置，默认值为 32K。

HA 服务端消息的传输一直以上述步骤循环运行，每次事件处理完成后等待 1s。

RocketMQ HA 主从同步机制就讲解到这里，其主要交互流程如图 7-7 所示。

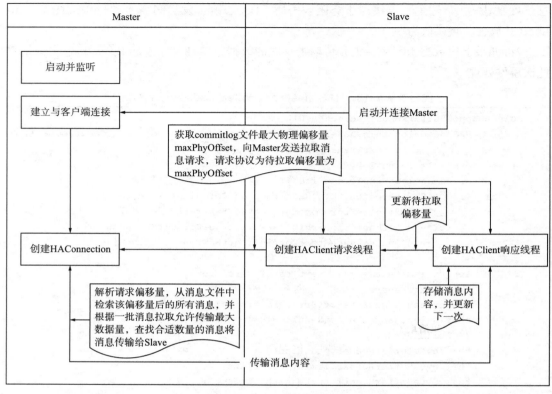

图 7-7 RocketMQ HA 交互类图

7.2 RocketMQ 读写分离机制

上节主要介绍了 RocketMQ 主从服务器的实现原理，本节主要介绍从服务器如何参与消息拉取负载机制。消息消费是基于消息消费队列 MessageQueue，再回顾一下 MessageQueue 队列的类图，如图 7-8 所示。

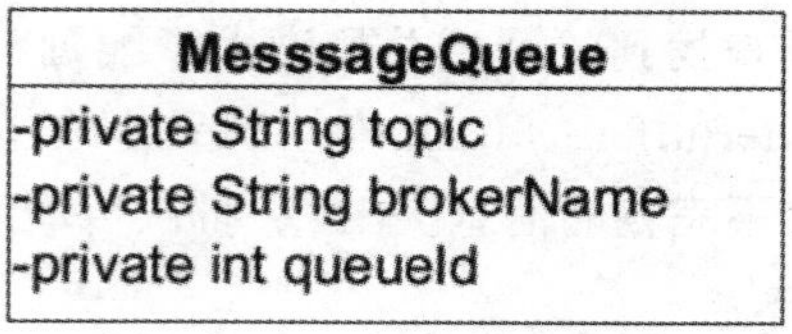

图 7-8 MessageQueue 类图

接下来将重点分析 RocketMQ 根据 brokerName 查找 Broker 地址的过程。
RocketMQ 根据 MessageQueue 查找 Broker 地址的唯一依据是 brokerName，从 Rocket-

MQ 的 Broker 组织结构中得知同一组 Broker（M-S）服务器，它们的 brokerName 相同但 brokerId 不同，主服务器的 brokerId 为 0，从服务器的 brokerId 大于 0，RocketMQ 提供 MQClientFactory.findBrokerAddressInSubscribe 来实现根据 brokerName、brokerId 查找 Broker 地址。

代码清单 7-19　PullAPIWrapper#pullKernelImpl

```
FindBrokerResult findBrokerResult =
    this.mQClientFactory.findBrokerAddressInSubscribe(mq.getBrokerName(),
    this.recalculatePullFromWhichNode(mq), false);
```

返回结果类图如图 7-9 所示。

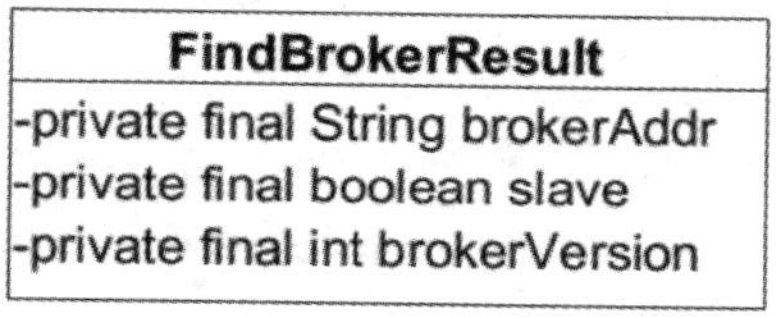

```
           FindBrokerResult
-private final String brokerAddr
-private final boolean slave
-private final int brokerVersion
```

图 7-9　FindBrokerResult 类图

代码清单 7-20　MQClientInstance#findBrokerAddressInSubscribe

```
public FindBrokerResult findBrokerAddressInSubscribe(
    String brokerName, long brokerId, boolean onlyThisBroker) {
    String brokerAddr = null;
    boolean slave = false;
    boolean found = false;
    HashMap<Long/* brokerId */, String/* address */> map =
        this.brokerAddrTable.get(brokerName);
    if (map != null && !map.isEmpty()) {
        brokerAddr = map.get(brokerId);
        slave = brokerId != MixAll.MASTER_ID;
        found = brokerAddr != null;
        if (!found && !onlyThisBroker) {
            Entry<Long, String> entry = map.entrySet().iterator().next();
            brokerAddr = entry.getValue();
            slave = entry.getKey() != MixAll.MASTER_ID;
            found = true;
        }
    }
    if (found) {
        return new FindBrokerResult(brokerAddr, slave, findBrokerVersion(brokerName,
            brokerAddr));
    }
    return null;
}
```

1）首先解释一下该方法的参数。
- brokerName：Broker 名称。
- brokerId：BrokerId。
- onlyThisBroker：是否必须返回 brokerId 的 Broker 对应的服务器信息。

2）从 ConcurrentMap<String/* Broker Name */, HashMap<Long/* brokerId */, String/* address */>> brokerAddrTable 地址缓存表中根据 brokerName 获取所有的 Broker 信息。

3）根据 brokerId 从 Broker 主从缓存表中获取指定 Broker 名称，如果根据 brokerId 未找到相关条目，此时若 onlyThisBroker 为 false，则随机返回 Broker 中任意一个 Broker，否则返回 null。

4）组装 FindBrokerResult 时，需要设置是否是 slave 这个属性。如果 brokerId=0，表示返回的 Broker 是主节点，否则返回的是从节点。

根据消息消费队列获取 brokerId 的实现如下。

代码清单 7-21　MQClientInstance#recalculatePullFromWhichNode

```
public long recalculatePullFromWhichNode(final MessageQueue mq) {
    if (this.isConnectBrokerByUser()) {
        return this.defaultBrokerId;
    }
    AtomicLong suggest = this.pullFromWhichNodeTable.get(mq);
    if (suggest != null) {
        return suggest.get();
    }
    return MixAll.MASTER_ID;
}
```

从 ConcurrentMap<MessageQueue, AtomicLong/* brokerId */> pullFromWhichNodeTable 缓存表中获取该消息消费队列的 brokerId，如果找到，则返回，否则返回 brokerName 的主节点。那 pullFromWhichNodeTable 消息从何而来呢？原来消息消费拉取线程 PullMessage-Service 根据 PullRequest 请求从主服务器拉取消息后会返回下一次建议拉取的 brokerId，消息消费者线程在收到消息后，会根据主服务器的建议拉取 brokerId 来更新 pullFromWhich-NodeTable，消息消费者线程更新 pullFromWhichNodeTable 的代码如下。

代码清单 7-22　PullAPIWrapper#processPullResult

```
this.updatePullFromWhichNode(mq, pullResultExt.getSuggestWhichBrokerId());
public void updatePullFromWhichNode(MessageQueue mq, long brokerId) {
    AtomicLong suggest = this.pullFromWhichNodeTable.get(mq);
    if (null == suggest) {
        this.pullFromWhichNodeTable.put(mq, new AtomicLong(brokerId));
    } else {
        suggest.set(brokerId);
```

}
}
```

那消息服务端是根据何种规则来建议哪个消息消费队列该从哪台 Broker 服务器上拉取消息呢？

**代码清单 7-23　DefaultMessageStore#getMessage**
```
long diff = maxOffsetPy - maxPhyOffsetPulling;
long memory = (long) (StoreUtil.TOTAL_PHYSICAL_MEMORY_SIZE *
        (this.messageStoreConfig.getAccessMessageInMemoryMaxRatio() / 100.0));
getResult.setSuggestPullingFromSlave(diff > memory);
```

1）maxOffsetPy：代表当前主服务器消息存储文件最大偏移量。

2）maxPhyOffsetPulling：此次拉取消息最大偏移量。

3）diff：对于 PullMessageService 线程来说，当前未被拉取到消息消费端的消息长度。

4）TOTAL_PHYSICAL_MEMORY_SIZE：RocketMQ 所在服务器总内存大小。accessMessageInMemoryMaxRatio 表示 RocketMQ 所能使用的最大内存比例，超过该内存，消息将被置换出内存；memory 表示 RocketMQ 消息常驻内存的大小，超过该大小，RocketMQ 会将旧的消息置换回磁盘。

5）如果 diff 大于 memory，表示当前需要拉取的消息已经超出了常驻内存的大小，表示主服务器繁忙，此时才建议从从服务器拉取。

**代码清单 7-24　PullMessageProcessor#processRequest**
```
if (getMessageResult.isSuggestPullingFromSlave()) {
    responseHeader.setSuggestWhichBrokerId(subscriptionGroupConfig.getWhichBrok
        erWhenConsumeSlowly());
} else {
    responseHeader.setSuggestWhichBrokerId(MixAll.MASTER_ID);
}
```

如果主服务器繁忙则建议下一次从从服务器拉取消息，设置 suggestWhichBrokerId 为配置文件中 whichBrokerWhenConsumeSlowly 属性，默认为 1。如果一个 Master 拥有多台 Slave 服务器，参与消息拉取负载的从服务器只会是其中一个。

## 7.3　本章小结

本章重点剖析了 RocketMQ HA 主从同步负载机制与主从服务器读写分离机制。

RocketMQ 的 HA 机制，其核心实现是从服务器在启动的时候主动向主服务器建立 TCP 长连接，然后获取服务器的 commitlog 最大偏移量，以此偏移量向主服务器主动拉取消息，主服务器根据偏移量，与自身 commitlog 文件的最大偏移量进行比较，如果大于从服务器
```

的 commitlog 偏移量，主服务器将向从服务器返回一定数量的消息，该过程循环进行，达到主从服务器数据同步。

RocketMQ 读写分离与其他中间件的实现方式完全不同，RocketMQ 是消费者首先向主服务器发起拉取消息请求，然后主服务器返回一批消息，然后会根据主服务器负载压力与主从同步情况，向从服务器建议下次消息拉取是从主服务器还是从从服务器拉取。

第 8 章 Chapter 8

RocketMQ 事务消息

最近 RocketMQ 官方发布了 4.3.0 版本，此版本解决了 RocketMQ 对事务的支持，这一重大更新对 RocketMQ 至关重要。本章将基于 RocketMQ 官方最新 4.3.0 版本，对其事务消息的实现原理进行深入探讨。主要内容如下。
- 事务消息实现思想
- 事务消息发送流程
- 事务消息提交或回滚
- 事务消息回查事务状态

8.1 事务消息实现思想

RocketMQ 事务消息的实现原理基于两阶段提交和定时事务状态回查来决定消息最终是提交还是回滚，交互设计如图 8-1 所示。

1）应用程序在事务内完成相关业务数据落库后，需要同步调用 RocketMQ 消息发送接口，发送状态为 prepare 的消息。消息发送成功后，RocketMQ 服务器会回调 RocketMQ 消息发送者的事件监听程序，记录消息的本地事务状态，该相关标记与本地业务操作同属一个事务，确保消息发送与本地事务的原子性。

2）RocketMQ 在收到类型为 prepare 的消息时，会首先备份消息的原主题与原消息消费队列，然后将消息存储在主题为 RMQ_SYS_TRANS_HALF_TOPIC 的消息消费队列中。

3）RocketMQ 消息服务器开启一个定时任务，消费 RMQ_SYS_TRANS_HALF_TOPIC 的消息，向消息发送端（应用程序）发起消息事务状态回查，应用程序根据保存的事务状态回馈消息服务器事务的状态（提交、回滚、未知），如果是提交或回滚，则消息服务器提交或回滚消息，如果是未知，待下一次回查，RocketMQ 允许设置一条消息的回查间隔与回查次数，如果在超过回查次数后依然无法获知消息的事务状态，则默认回滚消息。

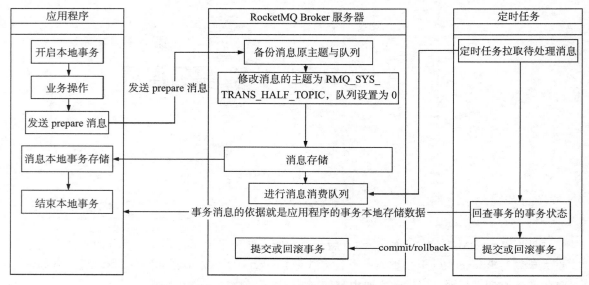

图 8-1　RocketMQ 事务消息实现原理

8.2　事务消息发送流程

RocketMQ 事务消息发送者为 org.apache.rocketmq.client.producer.TransactionMQProducer。其类继承图如图 8-2 所示。

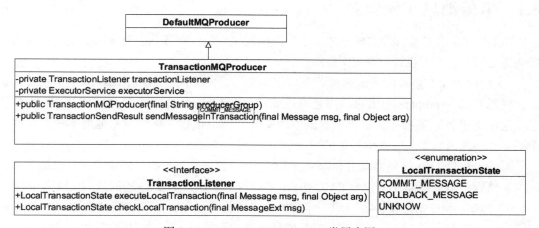

图 8-2　TransactionMQProducer 类层次图

其主要关键信息如下。

1. TransactionMQProducer

1）TransactionListener transactionListener：事务监听器，主要定义实现本地事务状态执行、本地事务状态回查两个接口。

2）ExecutorService executorService：事务状态回查异步执行线程池。

2. TransactionListener

1）LocalTransactionState executeLocalTransaction（final Message msg, final Object arg）：执行本地事务。

2）LocalTransactionState checkLocalTransaction（final MessageExt msg）：事务消息状态回查。

接下来详细分析事务消息发送流程。

代码清单 8-1　TransactionMQProducer#ssendMessageInTransaction

```
public TransactionSendResult sendMessageInTransaction(final Message msg, final
        Object arg) throws MQClientException {
    if (null == this.transactionListener) {
        throw new MQClientException("TransactionListener is null", null);
    }
    return this.defaultMQProducerImpl.sendMessageInTransaction(msg,
            transactionListener, arg);
}
```

如果事件监听器为空，则直接返回异常，最终调用 DefaultMQProducerImpl 的 sendMessageInTransaction 方法。

代码清单 8-2　DefaultMQProducerImpl#ssendMessageInTransaction

```
SendResult sendResult = null;
MessageAccessor.putProperty(msg, MessageConst.PROPERTY_TRANSACTION_PREPARED,
        "true");
MessageAccessor.putProperty(msg, MessageConst.PROPERTY_PRODUCER_GROUP,
        this.defaultMQProducer.getProducerGroup());
try {
    sendResult = this.send(msg);
} catch (Exception e) {
    throw new MQClientException("send message Exception", e);
}
```

Step1：首先为消息添加属性，TRAN_MSG 和 PGROUP，分别表示消息为 prepare 消息、消息所属消息生产者组。设置消息生产者组的目的是在查询事务消息本地事务状态时，从该生产者组中随机选择一个消息生产者即可，然后通过同步调用方式向 RocketMQ 发送消息，其发送消息的流程在第 2 章中有详细的分析，在本章稍后会重点关注针对事务消息所做的特殊处理。

代码清单 8-3　DefaultMQProducerImpl#ssendMessageInTransaction

```java
LocalTransactionState localTransactionState = LocalTransactionState.UNKNOW;
Throwable localException = null;
switch (sendResult.getSendStatus()) {
    case SEND_OK: {
        try {
            if (sendResult.getTransactionId() != null) {
                msg.putUserProperty("__transactionId__",
                    sendResult.getTransactionId());
            }
            String transactionId = msg.getProperty(MessageConst.PROPERTY_UNIQ_
                CLIENT_MESSAGE_ID_KEYIDX);
            if (null != transactionId && !"".equals(transactionId)) {
                msg.setTransactionId(transactionId);
            }
            localTransactionState = tranExecuter.executeLocalTransaction(msg, arg);
            if (null == localTransactionState) {
                localTransactionState = LocalTransactionState.UNKNOW;
            }
            if (localTransactionState != LocalTransactionState.COMMIT_MESSAGE) {
                log.info("executeLocalTransactionBranch return {}",
                    localTransactionState);
                log.info(msg.toString());
            }
        } catch (Throwable e) {
            log.info("executeLocalTransactionBranch exception", e);
            log.info(msg.toString());
            localException = e;
        }
    }
    break;
    case FLUSH_DISK_TIMEOUT:
    case FLUSH_SLAVE_TIMEOUT:
    case SLAVE_NOT_AVAILABLE:
        localTransactionState = LocalTransactionState.ROLLBACK_MESSAGE;
        break;
    default:
        break;
}
```

Step2：根据消息发送结果执行相应的操作。

1）如果消息发送成功，则执行 TransactionListener #executeLocalTransaction 方法，该方法的职责是记录事务消息的本地事务状态，例如可以通过将消息唯一 ID 存储在数据中，并且该方法与业务代码处于同一个事务，与业务事务要么一起成功，要么一起失败。这里是事务消息设计的关键理念之一，为后续的事务状态回查提供唯一依据。

2）如果消息发送失败，则设置本次事务状态为 LocalTransactionState.ROLLBACK_MESSAGE。

代码清单 8-4　DefaultMQProducerImpl#ssendMessageInTransaction

```
try {
    this.endTransaction(sendResult, localTransactionState, localException);
} catch (Exception e) {
    log.warn("local transaction execute " + localTransactionState + ", but end broker
        transaction failed", e);
}
```

Step3：结束事务。根据第二步返回的事务状态执行提交、回滚或暂时不处理事务。

1）LocalTransactionState.COMMIT_MESSAGE：提交事务。

2）LocalTransactionState.COMMIT_MESSAGE：回滚事务。

3）LocalTransactionState.UNKNOW：结束事务，但不做任何处理。

由于 this.endTransaction 的执行，其业务事务并没有提交，故在使用事务消息 TransactionListener #execute 方法除了记录事务消息状态后，应该返回 Local-Transaction.UNKNOW，事务消息的提交与回滚通过下面提到的事务消息状态回查时再决定是否提交或回滚。

事务消息发送的整体流程就介绍到这里了，接下来我们再重点介绍一下 prepare 消息发送的全过程。

代码清单 8-5　DefaultMQProducerImpl#sendKernelImpl

```
final String tranMsg =
    msg.getProperty(MessageConst.PROPERTY_TRANSACTION_PREPARED);
if (tranMsg != null && Boolean.parseBoolean(tranMsg)) {
    sysFlag |= MessageSysFlag.TRANSACTION_PREPARED_TYPE;
}
```

在消息发送之前，如果消息为 prepare 类型，则设置消息标准为 prepare 消息类型，方便消息服务器正确识别事务类型的消息。

代码清单 8-6　SendMessageProcessor#sendMessage

```
Map<String, String> oriProps = MessageDecoder.string2messageProperties
    (requestHeader.getProperties());
String traFlag = oriProps.get(MessageConst.PROPERTY_TRANSACTION_PREPARED);
if (traFlag != null && Boolean.parseBoolean(traFlag)) {
    if (this.brokerController.getBrokerConfig().isRejectTransactionMessage()) {
        response.setCode(ResponseCode.NO_PERMISSION);
        response.setRemark("the broker[" + this.brokerController.getBrokerConfig()
            .getBrokerIP1() + "] sending transaction message is forbidden");
        return response;
    }
```

```
            putMessageResult = this.brokerController.getTransactionalMessageService()
                .prepareMessage(msgInner);
        } else {
            putMessageResult = this.brokerController.getMessageStore().
                putMessage(msgInner);
        }
```

Broker 端在收到消息存储请求时，如果消息为 prepare 消息，则执行 prepareMessage 方法，否则走普通消息的存储流程。

代码清单 8-7　TransactionalMessageBridge#putHalfMessage

```
public PutMessageResult putHalfMessage(MessageExtBrokerInner messageInner) {
    return store.putMessage(parseHalfMessageInner(messageInner));
}
private MessageExtBrokerInner parseHalfMessageInner(MessageExtBrokerInner
        msgInner) {
    MessageAccessor.putProperty(msgInner, MessageConst.PROPERTY_REAL_TOPIC,
        msgInner.getTopic());
    MessageAccessor.putProperty(msgInner, MessageConst.PROPERTY_REAL_QUEUE_ID,
        String.valueOf(msgInner.getQueueId()));
    msgInner.setSysFlag(
        MessageSysFlag.resetTransactionValue(msgInner.getSysFlag(),
        MessageSysFlag.TRANSACTION_NOT_TYPE));
    msgInner.setTopic(TransactionalMessageUtil.buildHalfTopic());
    msgInner.setQueueId(0);
    msgInner.setPropertiesString(MessageDecoder.messageProperties2String(msgIn
        ner.getProperties()));
    return msgInner;
}
```

这里是事务消息与非事务消息发送流程的主要区别，如果是事务消息则备份消息的原主题与原消息消费队列，然后将主题变更为 RMQ_SYS_TRANS_HALF_TOPIC，消费队列变更为 0，然后消息按照普通消息存储在 commitlog 文件进而转发到 RMQ_SYS_TRANS_HALF_TOPIC 主题对应的消息消费队列。也就是说，事务消息在未提交之前并不会存入消息原有主题，自然也不会被消费者消费。既然变更了主题，RocketMQ 通常会采用定时任务（单独的线程）去消费该主题，然后将该消息在满足特定条件下恢复消息主题，进而被消费者消费。读者对这种实现应该并不陌生，它与 RocketMQ 定时消息的处理过程如出一辙。

RocketMQ 事务发送流程图如图 8-3 所示。

接下来重点分析一下调用结束事务 DefaultMQProducerImpl#endTransaction。

第 8 章 RocketMQ 事务消息

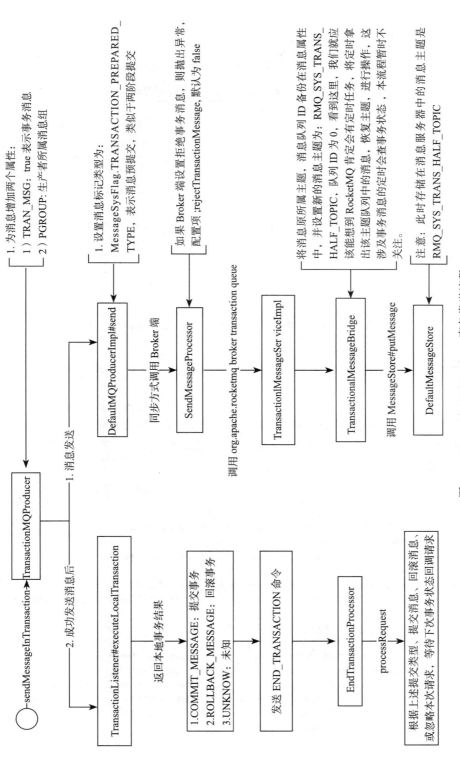

图 8-3 TransactionMQProducer 事务发送流程

8.3 提交或回滚事务

本节继续探讨两阶段提交的第二个阶段：提交或回滚事务。

代码清单 8-8　DefaultMQProducerImpl#endTransaction

```
String transactionId = sendResult.getTransactionId();
final String brokerAddr = this.mQClientFactory.findBrokerAddressInPublish
        (sendResult.getMessageQueue().getBrokerName());
EndTransactionRequestHeader requestHeader = new EndTransactionRequestHeader();
requestHeader.setTransactionId(transactionId);
requestHeader.setCommitLogOffset(id.getOffset());
switch (localTransactionState) {
case COMMIT_MESSAGE:
    requestHeader.setCommitOrRollback(MessageSysFlag.TRANSACTION_COMMIT_TYPE);
    break;
case ROLLBACK_MESSAGE:
    requestHeader.setCommitOrRollback(MessageSysFlag.
        TRANSACTION_ROLLBACK_TYPE);
    break;
case UNKNOW:
    requestHeader.setCommitOrRollback(MessageSysFlag.
        TRANSACTION_NOT_TYPE);
    break;
default:
    break;
}
```

根据消息所属的消息队列获取 Broker 的 IP 与端口信息，然后发送结束事务命令，其关键就是根据本地执行事务的状态分别发送提交、回滚或"不作为"的命令。Broker 服务端的结束事务处理器为：EndTransactionProcessor。

代码清单 8-9　EndTransactionProcessor#processRequest

```
OperationResult result = new OperationResult();
if (MessageSysFlag.TRANSACTION_COMMIT_TYPE == requestHeader.
        getCommitOrRollback()) {
    result = this.brokerController.getTransactionalMessageService().
        commitMessage(requestHeader);
    if (result.getResponseCode() == ResponseCode.SUCCESS) {
        RemotingCommand res = checkPrepareMessage(result.getPrepareMessage(),
            requestHeader);
        if (res.getCode() == ResponseCode.SUCCESS) {
            MessageExtBrokerInner msgInner =
                endMessageTransaction(result.getPrepareMessage());
            msgInner.setSysFlag(MessageSysFlag.resetTransactionValue
                (msgInner.getSysFlag(), requestHeader.getCommitOrRollback()));
            msgInner.setQueueOffset(requestHeader.getTranStateTableOffset());
            msgInner.setPreparedTransactionOffset(requestHeader.
                getCommitLogOffset());
```

```
            msgInner.setStoreTimestamp(result.getPrepareMessage().
               getStoreTimestamp());
            RemotingCommand sendResult = sendFinalMessage(msgInner);
            if (sendResult.getCode() == ResponseCode.SUCCESS) {
               this.brokerController.getTransactionalMessageService()
                  .deletePrepareMessage(result.getPrepareMessage());
            }
            return sendResult;
      }
      return res;
   }
}
```

如果结束事务动作为提交事务,则执行提交事务逻辑,其关键实现如下。

1)首先从结束事务请求命令中获取消息的物理偏移量(commitlogOffset),其实现逻辑由 TransactionalMessageService#.commitMessage 实现。

2)然后恢复消息的主题、消费队列,构建新的消息对象,由 TransactionalMessageService#endMessageTransaction 实现。

3)然后将消息再次存储在 commitlog 文件中,此时的消息主题则为业务方发送的消息,将被转发到对应的消息消费队列,供消息消费者消费,其实现由 TransactionalMessageService#sendFinalMessage 实现。

4)消息存储后,删除 prepare 消息,其实现方法并不是真正的删除,而是将 prepare 消息存储到 RMQ_SYS_TRANS_OP_HALF_TOPIC 主题中,表示该事务消息(prepare 状态的消息)已经处理过(提交或回滚),为未处理的事务进行事务回查提供查找依据。

事务的回滚与提交的唯一差别是无须将消息恢复原主题,直接删除 prepare 消息即可,同样是将预处理消息存储在 RMQ_SYS_TRANS_OP_HALF_TOPIC 主题中,表示已处理过该消息。

8.4 事务消息回查事务状态

上节重点梳理了 RocketMQ 基于两阶段协议发送与提交回滚消息,本节将深入学习事务状态消息回查,事务消息存储在消息服务器时主题被替换为 RMQ_SYS_TRANS_HALF_TOPIC,执行完本地事务返回本地事务状态为 UN_KNOW 时,结束事务时将不做任何处理,而是通过事务状态定时回查以期得到发送端明确的事务操作(提交事务或回滚事务)。

RocketMQ 通过 TransactionalMessageCheckService 线程定时去检测 RMQ_SYS_TRANS_HALF_TOPIC 主题中的消息,回查消息的事务状态。TransactionalMessageCheckService 的检测频率默认为 1 分钟,可通过在 broker.conf 文件中设置 transactionCheckInterval 来改变默认值,单位为毫秒。

代码清单 8-10　TransactionalMessageCheckService#onWaitEnd

```
protected void onWaitEnd() {
    long timeout = brokerController.getBrokerConfig().getTransactionTimeOut();
    int checkMax = brokerController.getBrokerConfig().getTransactionCheckMax();
    long begin = System.currentTimeMillis();
    log.info("Begin to check prepare message, begin time:{}", begin);
    this.brokerController.getTransactionalMessageService().check(timeout,
        checkMax, this.brokerController.getTransactionalMessageCheckListener());
    log.info("End to check prepare message, consumed time:{}",
            System.currentTimeMillis() - begin);
}
```

transactionTimeOut：事务的过期时间，只有当消息的存储时间加上过期时间大于系统当前时间时，才对消息执行事务状态回查，否则在下一次周期中执行事务回查操作。

transactionCheckMax：事务回查最大检测次数，如果超过最大检测次数还是无法获知消息的事务状态，RocketMQ 将不会继续对消息进行事务状态回查，而是直接丢弃即相当于回滚事务。

接下来重点分析 TransactionalMessageService#check 的实现逻辑，其实现类为 org.apache.rocketmq.broker.transaction.queue.TransactionalMessageServiceImpl。

代码清单 8-11　TransactionalMessageServiceImpl#check

```
String topic = MixAll.RMQ_SYS_TRANS_HALF_TOPIC;
Set<MessageQueue> msgQueues =
        transactionalMessageBridge.fetchMessageQueues(topic);
if (msgQueues == null || msgQueues.size() == 0) {
    log.warn("The queue of topic is empty :" + topic);
    return;
}
```

获取 RMQ_SYS_TRANS_HALF_TOPIC 主题下的所有消息队列，然后依次处理。

代码清单 8-12　TransactionalMessageServiceImpl#check

```
long startTime = System.currentTimeMillis();
MessageQueue opQueue = getOpQueue(messageQueue);
long halfOffset = transactionalMessageBridge.fetchConsumeOffset(messageQueue);
long opOffset = transactionalMessageBridge.fetchConsumeOffset(opQueue);
log.info("Before check, the queue={} msgOffset={} opOffset={}", messageQueue,
        halfOffset, opOffset);
if (halfOffset < 0 || opOffset < 0) {
    log.error("MessageQueue: {} illegal offset read: {}, op offset: {},skip this
        queue", messageQueue,halfOffset, opOffset);
    continue;
}
```

根据事务消息消费队列获取与之对应的消息队列，其实就是获取已处理消息的消息消费队列，其主题为：RMQ_SYS_TRANS_OP_HALF_TOPIC。

第 8 章 RocketMQ 事务消息 ❖ 235

代码清单 8-13　TransactionalMessageServiceImpl#check

```
    List<Long> doneOpOffset = new ArrayList<>();
    HashMap<Long, Long> removeMap = new HashMap<>();
    PullResult pullResult = fillOpRemoveMap(removeMap, opQueue, opOffset,
            halfOffset, doneOpOffset);
    if (null == pullResult) {
        log.error("The queue={} check msgOffset={} with opOffset={} failed, pullResult
            is null",messageQueue, halfOffset, opOffset);
        continue;
    }
```

fillOpRemoveMap 主要的作用是根据当前的处理进度依次从已处理队列拉取 32 条消息，方便判断当前处理的消息是否已经处理过，如果处理过则无须再次发送事务状态回查请求，避免重复发送事务回查请求。事务消息的处理涉及如下两个主题。

- RMQ_SYS_TRANS_HALF_TOPIC：prepare 消息的主题，事务消息首先进入到该主题。
- RMQ_SYS_TRANS_OP_HALF_TOPIC：当消息服务器收到事务消息的提交或回滚请求后，会将消息存储在该主题下。

代码清单 8-14　TransactionalMessageServiceImpl#check

```
// single thread int getMessageNullCount = 1; long newOffset = halfOffset; long
i = halfOffset; // @1
while (true) {
    if(System.currentTimeMillis() - startTime > MAX_PROCESS_TIME_LIMIT) { // @2
        log.info("Queue={} process time reach max={}", messageQueue,
            MAX_PROCESS_TIME_LIMIT);
        break;
    }
    if (removeMap.containsKey(i)) { // @3
        log.info("Half offset {} has been committed/rolled back", i);
        removeMap.remove(i);
    } else {
        GetResult getResult = getHalfMsg(messageQueue, i); // @4
        MessageExt msgExt = getResult.getMsg();
        if (msgExt == null) { // @5
            if (getMessageNullCount++ > MAX_RETRY_COUNT_WHEN_HALF_NULL) {
                break;
            }
            if (getResult.getPullResult().getPullStatus() == PullStatus.NO_NEW_MSG) {
                log.info("No new msg, the miss offset={} in={}, continue check={}, pull
                    result={}", i, messageQueue, getMessageNullCount,
                    getResult.getPullResult());
                break;
            } else {
                log.info("Illegal offset, the miss offset={} in={}, continuecheck={}, pull
                    result={}", i, messageQueue, getMessageNullCount,
                    getResult.getPullResult());
```

```
                    i = getResult.getPullResult().getNextBeginOffset();
                    newOffset = i;
                    continue;
                }
            }
            if (needDiscard(msgExt, transactionCheckMax) || needSkip(msgExt)) { // @6
                listener.resolveDiscardMsg(msgExt);
                newOffset = i + 1;
                i++; continue;
            }
            if (msgExt.getStoreTimestamp() >= startTime) {
                log.info("Fresh stored. the miss offset={}, check it later, store={}", i,
                    new Date(msgExt.getStoreTimestamp()));
                break;
            }
            long valueOfCurrentMinusBorn = System.currentTimeMillis() -
                msgExt.getBornTimestamp(); // @7
            long checkImmunityTime = transactionTimeout;
            String checkImmunityTimeStr =
            msgExt.getUserProperty(MessageConst.PROPERTY_CHECK_IMMUNITY_TIME_
                IN_SECONDS);
            if (null != checkImmunityTimeStr) { // @8
                checkImmunityTime = getImmunityTime(checkImmunityTimeStr,
                    transactionTimeout);
                if (valueOfCurrentMinusBorn < checkImmunityTime) {
                        if (checkPrepareQueueOffset(removeMap, doneOpOffset, msgExt,
                            checkImmunityTime)) {
                        newOffset = i + 1;
                        i++;
                        continue;
                    }
                }
            } else { // @9
                if ((0 <= valueOfCurrentMinusBorn) && (valueOfCurrentMinusBorn <
                    checkImmunityTime)) {
                    log.info("New arrived, the miss offset={}, check it later
                        checkImmunity={}, born={}", i, checkImmunityTime, new
                        Date(msgExt.getBornTimestamp()));
                    break;
                }
            }
            List<MessageExt> opMsg = pullResult.getMsgFoundList();
            boolean isNeedCheck =
                (opMsg == null && valueOfCurrentMinusBorn > checkImmunityTime) ||
                (opMsg != null && (opMsg.get(opMsg.size() - 1).getBornTimestamp() -
                startTime > transactionTimeout)) || (valueOfCurrentMinusBorn <= -1);
                // @10
            if (isNeedCheck) {
                if (!putBackHalfMsgQueue(msgExt, i)) {
                    // @11 continue;
                }
                listener.resolveHalfMsg(msgExt);
            } else {
```

```
                pullResult = fillOpRemoveMap(removeMap, opQueue,
                    pullResult.getNextBeginOffset(), halfOffset, doneOpOffset); // @12
                log.info("The miss offset:{} in messageQueue:{} need to get more opMsg,
                    result is:{}", i, messageQueue, pullResult);
            continue;
            }
        }
        newOffset = i + 1;
        i++;
    }
    if (newOffset != halfOffset) { // @13
        transactionalMessageBridge.updateConsumeOffset(messageQueue, newOffset);
    }
    long newOpOffset = calculateOpOffset(doneOpOffset, opOffset);
    if (newOpOffset != opOffset) { // @14
        transactionalMessageBridge.updateConsumeOffset(opQueue, newOpOffset);
    }
```

上述代码虽然比较多，但确是构成事务消息回查的核心处理逻辑。

代码 @1：先解释几个局部变量的含义。

❏ getMessageNullCount：获取空消息的次数。

❏ newOffset：当前处理 RMQ_SYS_TRANS_HALF_TOPIC#queueId 的最新进度。

❏ i：当前处理消息的队列偏移量，其主题依然为 RMQ_SYS_TRANS_HALF_TOPIC。

代码 @2：这段代码大家应该并不陌生，RocketMQ 处理任务的一个通用处理逻辑就是为每个任务一次只分配某个固定时长，超过该时长则需等待下次任务调度。RocketMQ 为待检测主题 RMQ_SYS_TRANS_HALF_TOPIC 的每个队列做事务状态回查，一次最多不超过 60 秒，目前该值不可配置。

代码 @3：如果该消息已被处理，则继续处理下一条消息。

代码 @4：根据消息队列偏移量 i 从消费队列中获取消息。

代码 @5：从待处理任务队列中拉取消息，如果未拉取到消息，则根据允许重复次数进行操作，默认重试一次，目前不可配置。其具体实现如下。

1）如果超过重试次数，直接跳出，结束该消息队列的事务状态回查。

2）如果是由于没有新的消息而返回为空（拉取状态为：PullStatus.NO_NEW_MSG），则结束该消息队列的事务状态回查。

3）其他原因，则将偏移量 i 设置为：getResult.getPullResult().getNextBeginOffset()，重新拉取。

代码 @6：判断该消息是否需要 discard（吞没、丢弃、不处理）或 skip（跳过），其依据如下。

1）needDiscard 依据：如果该消息回查的次数超过允许的最大回查次数，则该消息将被丢弃，即事务消息提交失败，具体实现方式为每回查一次，在消息属性 TRANSACTION_

CHECK_TIMES 中增 1，默认最大回查次数为 5 次。

2）needSkip 依据：如果事务消息超过文件的过期时间，默认为 72 小时（具体请查看 RocketMQ 过期文件相关内容），则跳过该消息。

代码 @7：处理事务超时相关概念，先解释几个局部变量。

- valueOfCurrentMinusBorn：消息已存储的时间，为系统当前时间减去消息存储的时间戳。
- checkImmunityTime：立即检测事务消息的时间，其设计的意义是，应用程序在发送事务消息后，事务不会马上提交，该时间就是假设事务消息发送成功后，应用程序事务提交的时间，在这段时间内，RocketMQ 任务事务未提交，故不应该在这个时间段向应用程序发送回查请求。
- transactionTimeout：事务消息的超时时间，这个时间是从 OP 拉取的消息的最后一条消息的存储时间与 check 方法开始的时间，如果时间差超过了 transactionTimeout，就算时间小于 checkImmunityTime 时间，也发送事务回查指令。

MessageConst.PROPERTY_CHECK_IMMUNITY_TIME_IN_SECONDS：消息事务消息回查请求的最晚时间，单位为秒，指的是程序发送事务消息时，可以指定该事务消息的有效时间，只有在这个时间内收到回查消息才有效，默认为 null。

代码 @8：如果消息指定了事务消息过期时间属性（PROPERTY_CHECK_IMMUNITY_TIME_IN_SECONDS），如果当前时间已超过该值。

代码 @9：如果当前时间还未过（应用程序事务结束时间），则跳出本次处理，等下一次再试。

代码 @10：判断是否需要发送事务回查消息，具体逻辑如下。

1）如果操作队列（RMQ_SYS_TRANS_OP_HALF_TOPIC）中没有已处理消息并且已经超过应用程序事务结束时间即 transactionTimeOut 值。

2）如果操作队列不为空并且最后一条消息的存储时间已经超过 transactionTimeOut 值。

代码 @11：如果需要发送事务状态回查消息，则先将消息再次发送到 RMQ_SYS_TRANS_HALF_TOPIC 主题中，发送成功则返回 true，否则返回 false，这里还有一个实现关键点。

代码清单 8-15　TransactionalMessageServiceImpl#putBackHalfMsgQueue

```
private boolean putBackHalfMsgQueue(MessageExt msgExt, long offset) {
    PutMessageResult putMessageResult = putBackToHalfQueueReturnResult(msgExt);
    if (putMessageResult != null
        && putMessageResult.getPutMessageStatus() == PutMessageStatus.PUT_OK) {
        msgExt.setQueueOffset(
            putMessageResult.getAppendMessageResult().getLogicsOffset());
        msgExt.setCommitLogOffset(
            putMessageResult.getAppendMessageResult().getWroteOffset());
        msgExt.setMsgId(putMessageResult.getAppendMessageResult().getMsgId());
```

```
        return true;
    } else {
        // 省略 error 日志
        return false;
        }
}
```

在执行事务消息回查之前,竟然在此把该消息存储在 commitlog 文件,新的消息设置最新的物理偏移量。为什么需要这样处理呢?主要是因为下文的发送事务消息是异步处理的,无法立刻知道其处理结果,为了避免简化 prepare 消息队列和处理队列的消息消费进度处理,先存储,然后消费进度向前推动,重复发送的消息在事务回查之前会判断是否处理过。另外一个目的就是需要修改消息的检查次数,RocketMQ 的存储设计采用顺序写,去修改已存储的消息,其性能无法高性能。

代码 @11:发送具体的事务回查命令,使用线程池来异步发送回查消息,为了回查消费进度保存的简化,只要发送了回查消息,当前回查进度会向前推动,如果回查失败,上一步骤新增的消息将可以再次发送回查消息,那如果回查消息发送成功,会不会下一次又重复发送回查消息呢?这个可以根据 OP 队列中的消息来判断是否重复,如果回查消息发送成功并且消息服务器完成提交或回滚操作,这条消息会发送到 OP 队列中,然后首先会通过 fillOpRemoveMap 根据处理进度获取一批已处理的消息,来与消息判断是否重复,由于 fillopRemoveMap 一次只拉 32 条消息,那又如何保证一定能拉取到与当前消息的处理记录呢?其实就是通过代码 @10,如果此批消息最后一条未超过事务延迟消息,则继续拉取更多消息进行判断(@12)和(@14),OP 队列也会随着回查进度的推进而推进。

代码 @12:如果无法判断是否发送回查消息,则加载更多的已处理消息进行筛选。

代码 @13:保存(Prepare)消息队列的回查进度。

代码 @14:保存处理队列(OP)的进度。

上述讲解了 TransactionalMessageCheckService 回查定时线程的发送回查消息的整体流程与实现细节,接下来重点分析一下上述步骤 @11,通过异步方式发送消息回查的实现过程。

代码清单 8-16 AbstractTransactionalMessageCheckListener#sendCheckMessage

```
public void sendCheckMessage(MessageExt msgExt) throws Exception {
    CheckTransactionStateRequestHeader checkTransactionStateRequestHeader = new
        CheckTransactionStateRequestHeader();
    checkTransactionStateRequestHeader.setCommitLogOffset(msgExt.getCommitLogO
        ffset());
    checkTransactionStateRequestHeader.setOffsetMsgId(msgExt.getMsgId());
    checkTransactionStateRequestHeader.setMsgId(msgExt.getUserProperty(MessageC
        onst.PROPERTY_UNIQ_CLIENT_MESSAGE_ID_KEYIDX));
    checkTransactionStateRequestHeader.setTransactionId(checkTransactionStateR
        equestHeader.getMsgId());
    checkTransactionStateRequestHeader.setTranStateTableOffset(msgExt.getQueue
Offset());
```

```
msgExt.setTopic(msgExt.getUserProperty(MessageConst.PROPERTY_REAL_TOPIC));
msgExt.setQueueId(Integer.parseInt(msgExt.getUserProperty(MessageConst.PRO
PERTY_REAL_QUEUE_ID)));
msgExt.setStoreSize(0);
String groupId = msgExt.getProperty(MessageConst.PROPERTY_PRODUCER_GROUP);
Channel channel = brokerController.getProducerManager()
    .getAvaliableChannel(groupId);
if (channel != null) {
brokerController.getBroker2Client().checkProducerTransactionState(groupI
d, channel, checkTransactionStateRequestHeader, msgExt);
    } else {
        LOGGER.warn("Check transaction failed, channel is null. groupId={}",
groupId);
    }
}
```

首先构建事务状态回查请求消息，核心参数包含消息 offsetId、消息 ID（索引）、消息事务 ID、事务消息队列中的偏移量、消息主题、消息队列。然后根据消息的生产者组，从中随机选择一个消息发送者。最后向消息发送者发送事务回查命令。

事务回查命令的最终处理者为 ClientRemotingProssor 的 processRequest 方法，最终将任务提交到 TransactionMQProducer 的线程池中执行，最终调用应用程序实现的 TransactionListener 的 checkLocalTransaction 方法，返回事务状态。如果事务状态为 LocalTransactionState#COMMIT_MESSAGE，则向消息服务器发送提交事务消息命令；如果事务状态为 LocalTransactionState#ROLLBACK_MESSAGE，则向 Broker 服务器发送回滚事务操作；如果事务状态为 UNOWN，则服务端会忽略此次提交。

事务状态回查的流程图 8-4 所示。

8.5　本章小结

本章重点分析了 RocketMQ 事务消息的实现机制，一言以蔽之，RocketMQ 事务消息基于两阶段提交和事务状态回查机制来实现，所谓的两阶段提交，即首先发送 prepare 消息，待事务提交或回滚时发送 commit、rollback 命令。再结合定时任务，RocketMQ 使用专门的线程以特定的频率对 RocketMQ 服务器上的 prepare 信息进行处理，向发送端查询事务消息的状态来决定是否提交或回滚消息。

第 8 章 RocketMQ 事务消息

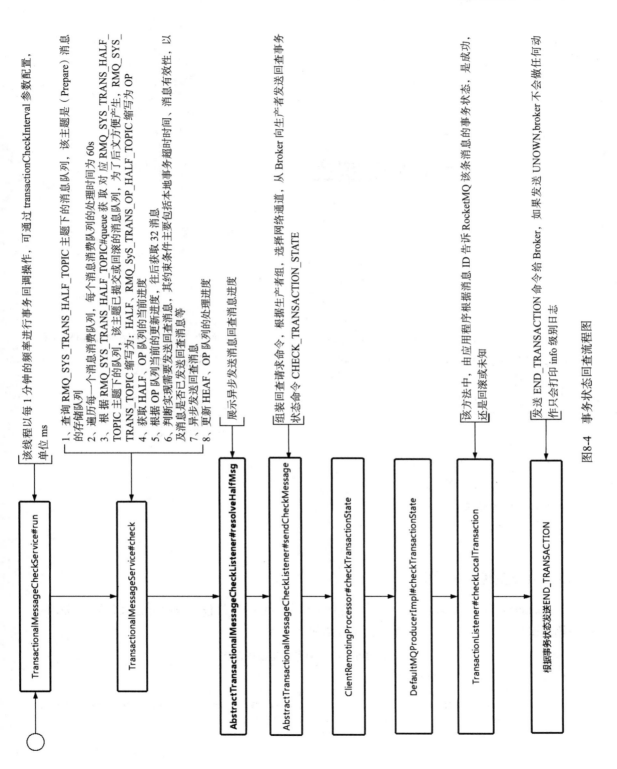

图 8-4 事务状态回查流程图

Chapter 9 第 9 章

RocketMQ 实战

前面介绍了 RocketMQ 的架构和运行原理等，本章主要介绍如何使用 RocketMQ 和 Spring 的整合、RocketMQ 和 Spring boot 等框架的整合。主要内容如下。

- 消息批量发送。
- 消息发送队列自选择。
- 消息过滤。
- 事务消息。
- Spring 整合 RocketMQ。
- Spring Cloud 整合 RocketMQ。
- RocketMQ 监控与运维命令。
- 应用场景分析。

9.1 消息批量发送

RocketMQ 消息批量发送是将同一主题的多条消息一起打包发送到消息服务端，减少网络调用次数，提高网络传输效率，使用方法如下。

代码清单 9-1 消息批量发送示例

```
public class SimpleBatchProducer {
    public static void main(String[] args) throws Exception {
        DefaultMQProducer producer=new DefaultMQProducer("BatchProducerGroupName");
        producer.setNamesrvAddr("127.0.0.1:9876");
        producer.start();
```

```
    String topic = "BatchTest";
    List<Message> messages = new ArrayList<>();
    messages.add(new Message(topic, "Tag", "OrderID001", "Hello world
        1".getBytes()));
    messages.add(new Message(topic, "Tag", "OrderID002", "Hello world
        2".getBytes()));
    messages.add(new Message(topic, "Tag", "OrderID003", "Hello world
        3".getBytes()));
    System.out.println(producer.send(messages));
    producer.shutdown();
    }
}
```

9.2 消息发送队列自选择

消息发送默认根据主题的路由信息（主题消息队列）进行负载均衡，负载均衡机制为轮询策略。例如现在有这样一个场景，订单的状态变更消息发送到特定主题，为了避免消息消费者同时消费同一订单的不同状态的变更消息，我们应该使用顺序消息。为了提高消息消费的并发度，如果我们能根据某种负载算法，相同订单的不同消息能统一发到同一个消息消费队列上，则可以避免引入分布式锁，RocketMQ 在消息发送时提供了消息队列选择器MessageQueueSelector。

代码清单 9-2　消息发送自定义分片算法

```
String[] tags = new String[] {"TagA", "TagB", "TagC", "TagD", "TagE"};
for (int i = 0; i < 100; i++) {
    int orderId = i % 10;
        Message msg = new Message("TopicTestjjj", tags[i % tags.length], "KEY" + i,
            ("Hello RocketMQ " + i).getBytes(RemotingHelper.DEFAULT_CHARSET));
        SendResult sendResult = producer.send(msg, new MessageQueueSelector() {
            public MessageQueue select(List<MessageQueue> mqs, Message msg, Object arg)
            {
                Integer id = (Integer) arg;
                int index = id % mqs.size();
                return mqs.get(index);
            }
        }, orderId);
        System.out.printf("%s%n", sendResult);
}
```

9.3 消息过滤

消息过滤包括基于表达式与基于类模式两种过滤模式，其中表达式又分为 TAG 与 SQL92 模式，接下来分别演示 3 种过滤模式的使用方法。

9.3.1　TAG 模式过滤

我们先来看下 TAG 模式过滤的例子。

代码清单 9-3　消息发送者发送待 TAG 的消息

```
for (int i = 0; i < 10; i++) {
    if( i % 2 == 0 ) {
    Message msg = new Message("TopicFilter7","TOPICA_TAG_ALL","OrderID001" ,
        "Helloworld".getBytes(RemotingHelper.DEFAULT_CHARSET));
    System.out.printf("%s%n", producer.send(msg));
    } else {
        Message msg = new Message("TopicFilter7","TOPICA_TAG_ORD","OrderID001",
            "Hello world".getBytes(RemotingHelper.DEFAULT_CHARSET));
        System.out.printf("%s%n", producer.send(msg));
    }
}
```

Step1：在消息发送时，我们可以为每一条消息设置一个 TAG 标签，消息消费者订阅自己感兴趣的 TAG，一般使用的场景是，对于同一类的功能（数据同步）创建一个主题，但对于该主题下的数据，可能不同的系统关心的数据不一样，但基础数据各个系统都需要同步，设置标签为 TOPICA_TAG_ALL，而订单数据只有订单下游子系统关心，其他系统并不关心，则设置标签为 TOPICA_TAG_ORD，库存子系统则关注库存相关的数据，设置标签为 TOPICA_TAG_CAPCITY。

代码清单 9-4　消息消费者订阅 TAG 模式

```
// 订单系统消费组
DefaultMQPushConsumer orderConsumer = new
    DefaultMQPushConsumer("Order_Data_Syn");
orderConsumer.subscribe("TopicFilter7", "TOPICA_TAG_ALL | TOPICA_TAG_ORD");
// 库存子系统消费组
DefaultMQPushConsumer kuCunConsumer = new
    DefaultMQPushConsumer("Order_Data_Syn");
kuCunConsumer.subscribe("TopicFilter7", "TOPICA_TAG_ALL | TOPICA_TAG_CAPCITY");
```

Step2：消费者组订阅相同的主题不同的 TAG，多个 TAG 用 "|" 分隔。注意：同一个消费组订阅的主题，TAG 必须相同。

9.3.2　SQL 表达模式过滤

对程序员来说，采用 SQL 表达式过滤是非常简单的，不用解释一看就明白，具体来看下代码。

代码清单 9-5　SQL 表达式消息发送方式

```
Message msg = new Message("TopicTest" /* Topic */,"TagA" /* Tag */,
    ("Hello RocketMQ " + i).getBytes(RemotingHelper.DEFAULT_CHARSET) );
msg.putUserProperty("orderStatus", "1");
```

```
msg.putUserProperty("sellerId", "21");
SendResult sendResult = producer.send(msg);
System.out.printf("%s%n", sendResult);
```

Step1：基于 SQL92 表达式消息过滤，其实是对消息的属性运用 SQL 过滤表达式进行条件匹配，所以消息发送时应该调用 putUserProperty 方法设置消息属性。

代码清单 9-6　基于 SQL 过滤消息消费者构建示例

```
consumer.subscribe("TopicTest",
    MessageSelector.bySql("(orderStatus is not null and orderStatus > 0 )"));
```

Step2：订阅模式为一条 SQL 条件过滤表达式，上下文环境为消息的属性。

9.3.3　类过滤模式

RocketMQ 通过定义消息过滤类的接口来实现自定义消息过滤，代码如下。

代码清单 9-7　org.apache.rocketmq.common.filter.MessageFilter 接口

```java
package org.apache.rocketmq.example.filter;
import org.apache.rocketmq.common.filter.FilterContext;
import org.apache.rocketmq.common.filter.MessageFilter;
import org.apache.rocketmq.common.message.MessageExt;
public class MessageFilterImpl implements MessageFilter {
    @Override
    public boolean match(MessageExt msg, FilterContext context) {
        String property = msg.getProperty("SequenceId");
        if (property != null) {
            int id = Integer.parseInt(property);
            if (((id % 10) == 0) &&
                (id > 100)) {
                return true;
            }
        }
        return false;
    }
}
```

Step1：实现自定义消息过滤器，实现 org.apache.rocketmq.common.filter.MessageFilter，MessageExt 实例中封装了整体消息的所有信息。

代码清单 9-8　自定义消息过滤类，实现 org.apache.rocketmq.common.filter.MessageFilter 接口

```java
DefaultMQPushConsumer consumer = new
    DefaultMQPushConsumer("ConsumerGroupNamecc4");
ClassLoader classLoader = Thread.currentThread().getContextClassLoader();
File classFile = new File(classLoader.getResource("MessageFilterImpl.java")
    .getFile());
```

```
String filterCode = MixAll.file2String(classFile);
consumer.subscribe("TopicTest",
    "org.apache.rocketmq.example.filter.MessageFilterImpl",filterCode);
```

Step2：消息消费者订阅主题，并上传自定义订阅类源码。

使用类过滤模式的前提是启动 FilterServer。

下面给出 Eclipse 调试 FilterServer 的方法，与在 Linux 环境部署 FilterServer 原理相同。

Step1：从 distribution 模块中将 logback_filtersrv.xml 复制到 ${ROCKETMQ_HOME}/conf 下，并新增 filtersrv.properites 文件，内容如下。

代码清单 9-9　filtersrv.properties 文件内容

```
#nameServer 地址，分号分割
namesrvAddr=127.0.0.1:9876
connectWhichBroker=127.0.0.1:10911
```

Step2：展开 filterSrv 模块，右键 FiltersrvStartup.java，移动到 Debug As，选中 Debug Configurations，切换到 Arguments 选项卡，增加 -c 配置选项，指定 FilterServer 配置文件，更多 FilterServer 配置文件属性请参考附录。操作对话框如图 9-1 所示。

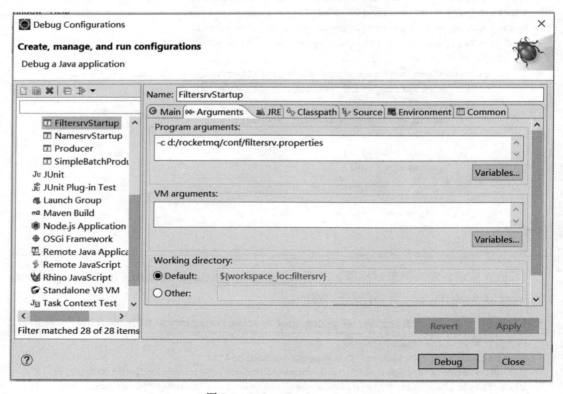

图 9-1　Debug Configurations

Step3：切换到 Environment 选项卡，配置 FilterServer 运行主目录，如图 9-2 所示。

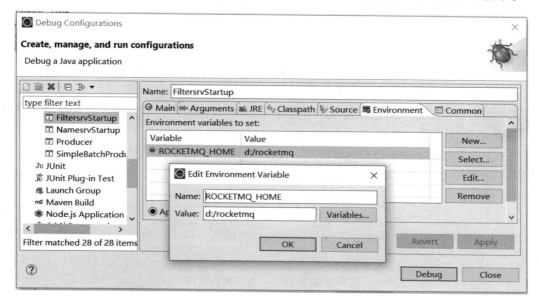

图 9-2　Debug Configurations

Step4：启动 FiltersrvStartup，如果控制台输出如下日志消息，表示启动成功，如果启动不成功，可以到 logback_broker.xml 中配置的日志文件查看错误日志。

代码清单 9-10　FilterServer 启动成功日志

```
load config properties file OK, d:/rocketmq/conf/filtersrv.properties
The Filter Server boot success, 192.168.1.3:62832
```

9.4　事务消息

我们以一个订单流转流程来举例，例如订单子系统创建订单，需要将订单数据下发到其他子系统（与第三方系统对接）这个场景。我们通常会将两个系统进行解耦，不直接使用服务调用的方式进行交互。其业务实现步骤通常有下面几步。

1）A 系统创建订单并入库。
2）发送消息到 MQ。
3）MQ 消费者消费消息，发送远程 RPC 服务调用，完成订单数据的同步。
方案一：

代码清单 9-11　订单下单伪代码

```
@SuppressWarnings("rawtypes")
public Map createOrder() {
```

```java
Map result = new HashMap();
// 执行下订单相关的业务流程，例如操作本地数据库落库相关代码
// 调用消息发送端 API 发送消息
// 返回结果，提交事务
return result;
}
```

方案一有以下弊端。

1）如果消息发送成功，在提交事务的时候 JVM 突然挂掉，事务没有成功提交，导致两个系统之间数据不一致。

2）由于消息是在事务提交之前提交，发送的消息内容是订单实体的内容，会造成在消费端进行消费时如果需要验证订单是否存在时可能出现订单不存在的情况。

3）消息发送可以考虑异步发送。

方案二：

由于存在上述问题，在 MQ 不支持事务消息的前提条件下，可以采用下面的方式进行优化。

代码清单 9-12　订单下单伪代码

```java
@SuppressWarnings("rawtypes")
public Map createOrder() {
    Map result = new HashMap();
    // 执行下订单相关的业务流程，例如操作本地数据库落库相关代码
    // 生成事务消息唯一业务表示，将该业务表示组装到待发送的消息体中
    // 往待发送消息表中插入一条记录，本次唯一消息发送业务 ID，消息 JSON{消息主题、消息 tag、消
    // 息体 }、创建时间、发送状态
    // 将消息体返回到控制器层
    // 返回结果，提交事务
    return result;
}
```

然后在控制器层异步发送消息，同时需要引入定时机制，去扫描待发送消息记录，避免消息丢失。

方案二有以下弊端。

1）消息有可能重复发送，但在消费端可以通过唯一业务编号来进行去重设计。

2）实现过于复杂，为了避免极端情况下的消息丢失，需要使用定时任务。

方案三：基于 RocketMQ4.3 版本事务消息。

代码清单 9-13　订单下单伪代码

```java
@SuppressWarnings("rawtypes")
public Map createOrder() {
    Map result = new HashMap();
    // 执行下订单相关的业务流程，例如操作本地数据库落库相关代码
    // 生成事务消息唯一业务表示，将该业务表示组装到待发送的消息体中，方便消息消费端进行幂等消费。
    // 调用消息客户端 API，发送事务 prepare 消息消费。
```

```
        // 返回结果,提交事务
        return result;
    }
```

上述是第一步,发送事务消息,接下来需要实现 TransactionListener,实现执行本地事务与本地事务回查。

代码清单 9-14 TransactionListener 监听器实现伪代码

```
import org.apache.rocketmq.client.producer.LocalTransactionState;
import org.apache.rocketmq.client.producer.TransactionListener; import org.apache.rocketmq.common.message.Message;
import org.apache.rocketmq.common.message.MessageExt;
import java.util.concurrent.ConcurrentHashMap;
@SuppressWarnings("unused")
public class OrderTransactionListenerImpl implements TransactionListener {
    private ConcurrentHashMap<String, Integer> countHashMap = new
        ConcurrentHashMap<>();
    private final static int MAX_COUNT = 5;
    @Override
    public LocalTransactionState executeLocalTransaction(Message msg, Object arg) {
        String bizUniNo = msg.getUserProperty("bizUniNo");// 从消息中获取业务唯一 ID。
        // 将 bizUniNo 入库,表名: t_message_transaction,表结构 bizUniNo(主键),业务类型。
        return LocalTransactionState.UNKNOW;
    }
    @Override
    public LocalTransactionState checkLocalTransaction(MessageExt msg) {
        Integer status = 0;
        // 从数据库查询 t_message_transaction 表,如果该表中存在记录,则提交,
        String bizUniNo = msg.getUserProperty("bizUniNo"); // 从消息中获取业务唯一 ID。
         // 然后查询 t_message_transaction  表,是否存在 bizUniNo,如果存在,则返回 COMMIT_MESSAGE,
        // 不存在,则记录查询次数,未超过次数,返回 UNKNOW,超过次数,返回 ROLLBACK_MESSAGE
        if(query(bizUniNo) > 0 ) {
            return LocalTransactionState.ROLLBACK_MESSAGE;
        }
        return rollBackOrUnown(bizUniNo);
    }
    public int query(String bizUniNo) {
        return 1; //select count(1) from t_message_transaction a where
            //a.biz_uni_no=#{bizUniNo}
    }
    public LocalTransactionState rollBackOrUnown(String bizUniNo) {
        Integer num = countHashMap.get(bizUniNo);
        if(num != null && ++num > MAX_COUNT) {
            countHashMap.remove(bizUniNo);
            return LocalTransactionState.ROLLBACK_MESSAGE;
        }
        if(num == null) {
```

```
            num = new Integer(1);
        }
        countHashMap.put(bizUniNo, num);
        return LocalTransactionState.UNKNOW;
    }
}
```

TransactionListener 实现要点如下。

❏ executeLocalTransaction：该方法主要是设置本地事务状态，与业务方代码在一个事务中，例如 OrderServer#createMap 中，只要本地事务提交成功，该方法也会提交成功。故在这里，主要是向 t_message_transaction 添加一条记录，在事务回查时，如果存在记录，就认为是该消息需要提交，其返回值建议返还 LocalTransactionState.UNKNOW。

❏ checkLocalTransaction：该方法主要是告知 RocketMQ 消息是需要提交还是回滚，如果本地事务表（t_message_transaction）存在记录，则认为提交；如果不存在，可以设置回查次数，如果指定次数内还是未查到消息，则回滚，否则返回未知。rocketmq 会按一定的频率回查事务，当然回查次数也有限制，默认为 5 次，可配置。

9.5 Spring 整合 RocketMQ

Spring 整合 RocketMQ 非常简单，即将 Spring 作为 RocketMQ 消息消费者、消息生产者的 Bean 容器。

代码清单 9-15　配置文件

```xml
<bean id = "mqProducer"
        class="persistent.prestige.web.base.mq.rocketmq.MqProducer"
        init-method="init" destroy-method="destroy">
    <property name="producerGroup"><value>MyProducerGroup</value></property>
    <property name="namesrvAddr">
        <value>192.168.56.166:9876</value>
    </property>
</bean>
<bean id = "mqConsume"
        class="persistent.prestige.web.base.mq.rocketmq.MqConsume"
        init-method="init" destroy-method="destroy">
    <property name="consumerGroup">
        <value>MyConsumerGroup</value>
    </property>
    <property name="namesrvAddr">
        <value>192.168.56.166:9876</value>
    </property>
</bean>
```

9.6 Spring Cloud 整合 RocketMQ

Step1：新建 project 以及添加依赖包。

代码清单 9-16　POM 文件

```xml
<?xml version="1.0" encoding="UTF-8"?>
<project xmlns="http://maven.apache.org/POM/4.0.0"
         xmlns:xsi="http://www.w3.org/2001/XMLSchema-instance"
         xsi:schemaLocation="http://maven.apache.org/POM/4.0.0 http://maven.apache.org/xsd/maven-4.0.0.xsd">
    <modelVersion>4.0.0</modelVersion>
    <groupId>com.rocketmq</groupId>
    <artifactId>test-rocketmq</artifactId>
    <version>1.0-SNAPSHOT</version>
    <packaging>jar</packaging>

    <name>test-rocketmq</name>
    <url>http://maven.apache.org</url>
    <description>RocketMQ</description>
    <properties>
        <project.build.sourceEncoding>UTF-8</project.build.sourceEncoding>
        <java_source_version>1.8</java_source_version>
        <java_target_version>1.8</java_target_version>
        <file_encoding>UTF-8</file_encoding>

        <springboot.version>1.5.9.RELEASE</springboot.version>
        <rocketmq.version>4.2.0</rocketmq.version>
    </properties>
    <dependencies>
        <!--SpringBoot-->
        <dependency>
            <groupId>org.springframework.boot</groupId>
            <artifactId>spring-boot-starter</artifactId>
            <version>${springboot.version}</version>
        </dependency>
        <dependency>
            <groupId>org.springframework.boot</groupId>
            <artifactId>spring-boot-starter-test</artifactId>
            <version>${springboot.version}</version>
            <scope>test</scope>
        </dependency>
        <!--Rocketmq-->
        <dependency>
            <groupId>org.apache.rocketmq</groupId>
            <artifactId>rocketmq-client</artifactId>
            <version>${rocketmq.version}</version>
        </dependency>
    </dependencies>
    <build>
        <plugins>
```

```xml
        <plugin>
            <groupId>org.apache.maven.plugins</groupId>
            <artifactId>maven-compiler-plugin</artifactId>
            <configuration>
                <fork>true</fork>
                <source>${java_source_version}</source>
                <target>${java_target_version}</target>
                <encoding>${file_encoding}</encoding>
            </configuration>
        </plugin>
      </plugins>
   </build>
</project>
```

Step2：Spring boot 启动类非常简单，只要加上 @Configuration 和 @EnableAutoConfiguration 注解即可。

代码清单 9-17　Spring boot 启动类

```java
package com.rocketmq.test;

import org.springframework.boot.SpringApplication;
import org.springframework.boot.autoconfigure.EnableAutoConfiguration;
import org.springframework.context.annotation.ComponentScan;
import org.springframework.context.annotation.Configuration;
@Configuration
@EnableAutoConfiguration
@ComponentScan(basePackages = {" com.rocketmq.test "})
public class RocketMQApplicationMain {

    public static void main(String[] args) {
        SpringApplication.run(RocketMQApplicationMain.class, args);
    }

}
```

Spring boot 启动还需要 application 的配置文件，配置的内容如下。

```yaml
spring:
   rocketmq:
       namesrvaddr: localhost:29876
       producerGroup:TestProducer
       consumerGroup:TestConsumer
```

Step3：生产者和消费者的代码。

代码清单 9-18　生产者代码

```java
package com.rocketmq.test;

import org.apache.rocketmq.client.exception.MQClientException;
import org.apache.rocketmq.client.producer.DefaultMQProducer;
```

```java
import org.apache.rocketmq.client.producer.SendCallback;
import org.apache.rocketmq.client.producer.SendResult;
import org.apache.rocketmq.common.message.Message;
import org.apache.rocketmq.remoting.common.RemotingHelper;
import org.slf4j.Logger;
import org.slf4j.LoggerFactory;
import org.springframework.beans.factory.annotation.Value;
import org.springframework.stereotype.Component;

import javax.annotation.PostConstruct;
import javax.annotation.PreDestroy;

/**
 * RocketMQ 生产者
 */
@Component
public class MQProducer {

    private static final Logger LOGGER =
LoggerFactory.getLogger(MQProducer.class);

    @Value("${spring.rocketmq.namesrvAddr}")
    private String namesrvAddr;

    private final DefaultMQProducer producer = new
DefaultMQProducer("TestRocketMQProducer");

    /**
     * 初始化
     */
    @PostConstruct
    public void start() {
        try {
            LOGGER.info("MQ:启动生产者");
            producer.setNamesrvAddr(namesrvAddr);
            producer.start();
        } catch (MQClientException e) {
            LOGGER.error("MQ:启动生产者失败: {}-{}", e.getResponseCode(),
e.getErrorMessage());
            throw new RuntimeException(e.getMessage(), e);
        }
    }

    /**
     * 发送消息
     *
     * @param data  消息内容
     * @param topic 主题
     * @param tags  标签
     * @param keys  唯一主键
     */
```

```java
    public void sendMessage(String data, String topic, String tags, String keys)
{
        try {
            byte[] messageBody = data.getBytes(RemotingHelper.DEFAULT_CHARSET);

            Message mqMsg = new Message(topic, tags, keys, messageBody);

            producer.send(mqMsg, new SendCallback() {
                @Override
                public void onSuccess(SendResult sendResult) {
                    LOGGER.info("MQ：生产者发送消息 {}", sendResult);
                }

                @Override
                public void onException(Throwable throwable) {
                    LOGGER.error(throwable.getMessage(), throwable);
                }
            });
        } catch (Exception e) {
            LOGGER.error(e.getMessage(), e);
        }

    }

    @PreDestroy
    public void stop() {
        if (producer != null) {
            producer.shutdown();
            LOGGER.info("MQ：关闭生产者 ");
        }
    }
}
```

代码清单9-19 消费者代码

```java
package com.rocketmq.test;

import org.apache.rocketmq.client.consumer.DefaultMQPushConsumer;
import org.apache.rocketmq.client.consumer.listener.ConsumeConcurrentlyContext;
import org.apache.rocketmq.client.consumer.listener.ConsumeConcurrentlyStatus;
import org.apache.rocketmq.client.consumer.listener.MessageListenerConcurrently;
import org.apache.rocketmq.client.exception.MQClientException;
import org.apache.rocketmq.common.consumer.ConsumeFromWhere;
import org.apache.rocketmq.common.message.MessageExt;
import org.apache.rocketmq.common.protocol.heartbeat.MessageModel;
import org.apache.rocketmq.remoting.common.RemotingHelper;
import org.slf4j.Logger;
import org.slf4j.LoggerFactory;
import org.springframework.beans.factory.annotation.Value;
import org.springframework.stereotype.Component;

import javax.annotation.PostConstruct;
import javax.annotation.PreDestroy;
import java.util.List;

/**
```

```java
 * RocketMQ 消费者
 */
@Component
public class MQPushConsumer implements MessageListenerConcurrently {

    private static final Logger LOGGER = LoggerFactory.getLogger(MQPushConsumer.class);

    @Value("${spring.rocketmq.namesrvAddr}")
    private String namesrvAddr;

    private final DefaultMQPushConsumer consumer = new DefaultMQPushConsumer("TestRocketMQPushConsumer");

    /**
     * 初始化
     *
     * @throws MQClientException
     */
    @PostConstruct
    public void start() {
        try {
            LOGGER.info("MQ:启动消费者 ");

            consumer.setNamesrvAddr(namesrvAddr);
            // 从消息队列头开始消费
            consumer.setConsumeFromWhere(ConsumeFromWhere.CONSUME_FROM_FIRST_OFFSET);
            // 集群消费模式
            consumer.setMessageModel(MessageModel.CLUSTERING);
            // 订阅主题
            consumer.subscribe("TopicTest", "*");
            // 注册消息监听器
            consumer.registerMessageListener(this);
            // 启动消费端
            consumer.start();
        } catch (MQClientException e) {
            LOGGER.error("MQ:启动消费者失败: {}-{}", e.getResponseCode(), e.getErrorMessage());
            throw new RuntimeException(e.getMessage(), e);
        }

    }

    /**
     * 消费消息
     * @param msgs
     * @param context
     * @return
     */
    @Override
    public ConsumeConcurrentlyStatus consumeMessage(List<MessageExt> msgs, ConsumeConcurrentlyContext context) {
        int index = 0;
        try {
```

```java
                for (; index < msgs.size(); index++) {
                    MessageExt msg = msgs.get(index);
                    String messageBody = new String(msg.getBody(),
RemotingHelper.DEFAULT_CHARSET);
                    LOGGER.info("MQ：消费者接收新信息：{} {} {} {} {}", msg.getMsgId(),
msg.getTopic(), msg.getTags(), msg.getKeys(), messageBody);
                }
            } catch (Exception e) {
                LOGGER.error(e.getMessage(), e);
            } finally {
                if (index < msgs.size()) {
                    context.setAckIndex(index + 1);
                }
            }
            return ConsumeConcurrentlyStatus.CONSUME_SUCCESS;
        }

        @PreDestroy
        public void stop() {
            if (consumer != null) {
                consumer.shutdown();
                LOGGER.error("MQ：关闭消费者");
            }
        }

}
```

Step4：测试的代码。

代码清单 9-20　测试的代码

```java
package com.rocketmq.test;

import com.xusg.study.producer.MQProducer;
import org.junit.Test;
import org.junit.runner.RunWith;
import org.springframework.boot.test.context.SpringBootTest;
import org.springframework.test.context.junit4.SpringRunner;

import javax.annotation.Resource;

@RunWith(SpringRunner.class)
@SpringBootTest(classes = {RocketMQApplicationMain.class})
public class TestRocketMQ {

    @Resource
    private MQProducer mqProducer;

    @Test
    public void testProducer() {
        for (int i = 0; i < 10; i++) {
```

```java
            mqProducer.sendMessage("Hello RocketMQ " + i, "TopicTest",
                    "TagTest", "Key" + i);
        }

        try {
            Thread.sleep(10 * 1000);
        } catch (InterruptedException e) {
            e.printStackTrace();
        }
    }
}
```

Step5：测试的结果。

代码清单 9-21　测试的结果

```
# 控制台日志信息
[02-06 15:08:52 500] main INFO MQPushConsumer - MQ：启动消费者
[02-06 15:08:53 622] main INFO MQProducer - MQ：启动生产者
[02-06 15:08:53 988] NettyClientPublicExecutor_1 INFO MQProducer - MQ：生产者发送
消息 SendResult [sendStatus=SEND_OK, msgId=C0A81E55346018B4AAC21CFFEB400001, offset
MsgId=C0A8386500002A9F00000000000814BA, messageQueue=MessageQueue [topic=TopicTest,
brokerName=broker-a, queueId=2], queueOffset=681]
[02-06 15:08:53 989] NettyClientPublicExecutor_1 INFO MQProducer - MQ：生产者发送
消息 SendResult [sendStatus=SEND_OK, msgId=C0A81E55346018B4AAC21CFFEB400003, offset
MsgId=C0A8386500002A9F0000000000081638, messageQueue=MessageQueue [topic=TopicTest,
brokerName=broker-a, queueId=0], queueOffset=678]
[02-06 15:08:53 989] ConsumeMessageThread_4 INFO MQPushConsumer - MQ：消费者接收新
信息：C0A81E55346018B4AAC21CFFEB3A0000 TopicTest TagTest Key0 Hello RocketMQ 0
[02-06 15:08:53 989] NettyClientPublicExecutor_1 INFO MQProducer - MQ：生产者发送
消息 SendResult [sendStatus=SEND_OK, msgId=C0A81E55346018B4AAC21CFFEB430004, offset
MsgId=C0A8386500002A9F00000000000816F7, messageQueue=MessageQueue [topic=TopicTest,
brokerName=broker-a, queueId=1], queueOffset=680]
[02-06 15:08:53 990] NettyClientPublicExecutor_1 INFO MQProducer - MQ：生产者发送
消息 SendResult [sendStatus=SEND_OK, msgId=C0A81E55346018B4AAC21CFFEB430005, offset
MsgId=C0A8386500002A9F00000000000817B6, messageQueue=MessageQueue [topic=TopicTest,
brokerName=broker-a, queueId=2], queueOffset=682]
[02-06 15:08:53 990] NettyClientPublicExecutor_1 INFO MQProducer - MQ：生产者发送
消息 SendResult [sendStatus=SEND_OK, msgId=C0A81E55346018B4AAC21CFFEB430006, offset
MsgId=C0A8386500002A9F0000000000081875, messageQueue=MessageQueue [topic=TopicTest,
brokerName=broker-a, queueId=3], queueOffset=682]
[02-06 15:08:53 990] NettyClientPublicExecutor_1 INFO MQProducer - MQ：生产者发送
消息 SendResult [sendStatus=SEND_OK, msgId=C0A81E55346018B4AAC21CFFEB430007, offset
MsgId=C0A8386500002A9F0000000000081934, messageQueue=MessageQueue [topic=TopicTest,
brokerName=broker-a, queueId=0], queueOffset=679]
[02-06 15:08:53 990] NettyClientPublicExecutor_1 INFO MQProducer - MQ：生产者发送
消息 SendResult [sendStatus=SEND_OK, msgId=C0A81E55346018B4AAC21CFFEB430008, offset
MsgId=C0A8386500002A9F00000000000819F3, messageQueue=MessageQueue [topic=TopicTest,
brokerName=broker-a, queueId=1], queueOffset=681]
[02-06 15:08:53 990] NettyClientPublicExecutor_1 INFO MQProducer - MQ：生产者发送
消息 SendResult [sendStatus=SEND_OK, msgId=C0A81E55346018B4AAC21CFFEB430009, offset
MsgId=C0A8386500002A9F000000000008 1AB2, messageQueue=MessageQueue [topic=TopicTest,
brokerName=broker-a, queueId=2], queueOffset=683]
```

```
    [02-06 15:08:53 990] NettyClientPublicExecutor_3 INFO MQProducer - MQ：生产者发送
消息 SendResult [sendStatus=SEND_OK, msgId=C0A81E55346018B4AAC21CFFEB400002, offset
MsgId=C0A8386500002A9F00000000000081579, messageQueue=MessageQueue [topic=TopicTest,
brokerName=broker-a, queueId=3], queueOffset=681]
    [02-06 15:08:53 990] NettyClientPublicExecutor_2 INFO MQProducer - MQ：生产者发送
消息 SendResult [sendStatus=SEND_OK, msgId=C0A81E55346018B4AAC21CFFEB3A0000, offset
MsgId=C0A8386500002A9F00000000000813FB, messageQueue=MessageQueue [topic=TopicTest,
brokerName=broker-a, queueId=1], queueOffset=679]
    [02-06 15:08:53 991] ConsumeMessageThread_1 INFO MQPushConsumer - MQ：消费者接收新
信息：C0A81E55346018B4AAC21CFFEB400001 TopicTest TagTest Key1 Hello RocketMQ 1
    [02-06 15:08:53 990] ConsumeMessageThread_3 INFO MQPushConsumer - MQ：消费者接收新
信息：C0A81E55346018B4AAC21CFFEB400002 TopicTest TagTest Key2 Hello RocketMQ 2
    [02-06 15:08:53 998] ConsumeMessageThread_2 INFO MQPushConsumer - MQ：消费者接收新
信息：C0A81E55346018B4AAC21CFFEB400003 TopicTest TagTest Key3 Hello RocketMQ 3
    [02-06 15:08:54 007] ConsumeMessageThread_9 INFO MQPushConsumer - MQ：消费者接收新
信息：C0A81E55346018B4AAC21CFFEB430008 TopicTest TagTest Key8 Hello RocketMQ 8
    [02-06 15:08:54 010] ConsumeMessageThread_5 INFO MQPushConsumer - MQ：消费者接收新
信息：C0A81E55346018B4AAC21CFFEB430005 TopicTest TagTest Key5 Hello RocketMQ 5
    [02-06 15:08:54 010] ConsumeMessageThread_6 INFO MQPushConsumer - MQ：消费者接收新
信息：C0A81E55346018B4AAC21CFFEB430009 TopicTest TagTest Key9 Hello RocketMQ 9
    [02-06 15:08:54 010] ConsumeMessageThread_7 INFO MQPushConsumer - MQ：消费者接收新
信息：C0A81E55346018B4AAC21CFFEB430006 TopicTest TagTest Key6 Hello RocketMQ 6
    [02-06 15:08:54 011] ConsumeMessageThread_8 INFO MQPushConsumer - MQ：消费者接收新
信息：C0A81E55346018B4AAC21CFFEB430004 TopicTest TagTest Key4 Hello RocketMQ 4
    [02-06 15:08:54 011] ConsumeMessageThread_10 INFO MQPushConsumer - MQ：消费者接收新
信息：C0A81E55346018B4AAC21CFFEB430007 TopicTest TagTest Key7 Hello RocketMQ 7
    [02-06 15:09:03 965] Thread-3 INFO MQProducer - MQ：关闭生产者
Disconnected from the target VM, address: '127.0.0.1:52407', transport: 'socket'
    [02-06 15:09:03 986] Thread-3 ERROR MQPushConsumer - MQ：关闭消费者
```

9.7 RocketMQ 监控与运维命令

一个完善的中间件，监控是必不可少的功能，通过监控我们可以查看系统是否运行正常，是否出现问题，是系统稳定性和运维的基础。

9.7.1 RocktetMQ 监控平台搭建

rocketmq-console 下载路径：https://github.com/875279177/incubator-rocketmq-externals，它是一个标准的 Maven 项目，基于 SpringBoot 并内嵌了 Web 服务器。

1）修改配置文件 application.properties，主要修改端口号、rocketmq.config.dataPath。

代码清单 9-22 rocketmq-console 配置文件

```
server.contextPath=
# web 服务器端口号
server.port=8080
#spring.application.index=true
spring.application.name=rocketmq-console
spring.http.encoding.charset=UTF-8
spring.http.encoding.enabled=true
```

```
spring.http.encoding.force=true
logging.config=classpath:logback.xml
#NameServer 地址
rocketmq.config.namesrvAddr=127.0.0.1:9876
rocketmq.config.isVIPChannel=
#rocketmq-console's data path:dashboard/monitor
rocketmq.config.dataPath=D:/rocketmq/logs/data
#set it false if you don't want use dashboard.default true
rocketmq.config.enableDashBoardCollect=true
```

2）日志文件，主要修改日志存放文件即可。

3）运行 org.apache.rocketmq.console.App 启动监控程式，在浏览器中输入 http://url:serverport 打开监控界面。

RocketMq 控制台的主界面非常简洁，有 6 个 Tab 页面。

1）Dashboard：监控图表主界面。

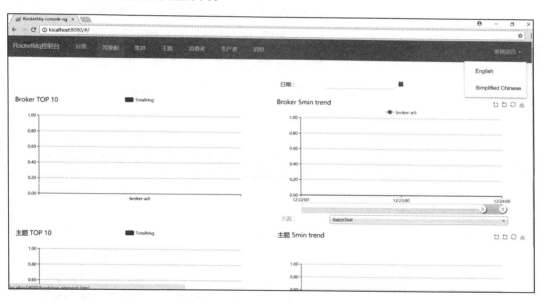

图 9-3　rocketmq-console Dashboard

2）Cluster：集群信息主界面。

图 9-4　rocketmq-console Cluster

3）Topic：主题管理界面。

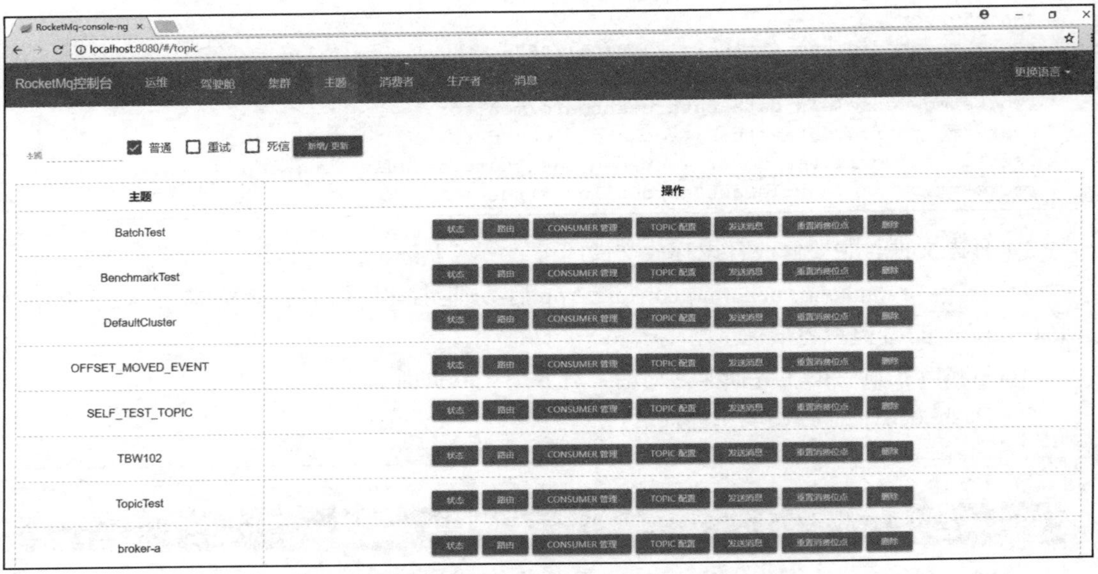

图 9-5　rocketmq-console Topic 界面

4）Consumer：消息消费主界面。

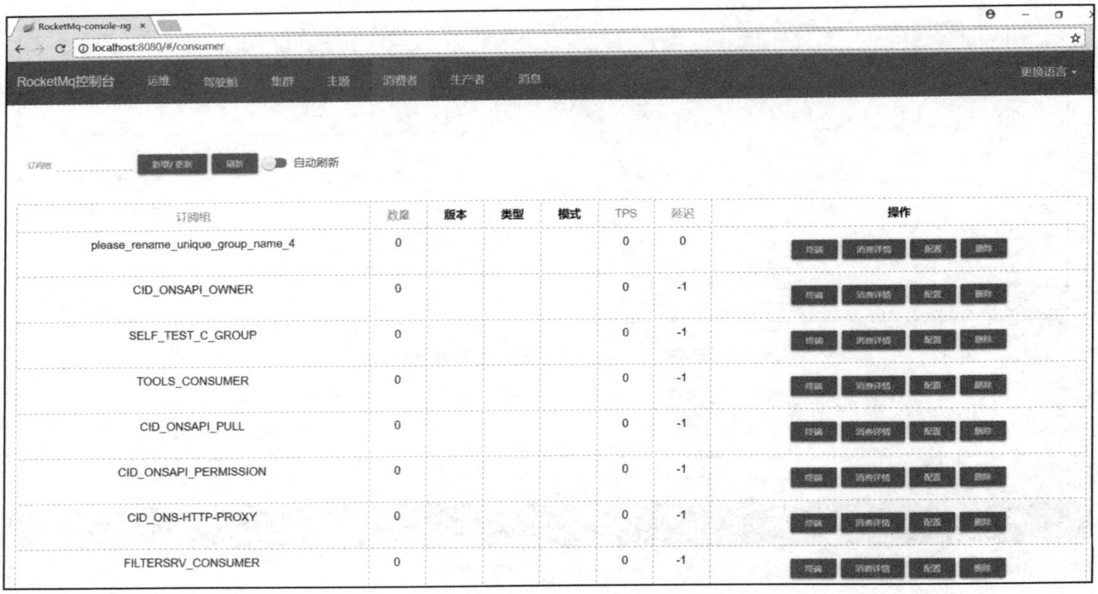

图 9-6　rocketmq-console Consumer 界面

5）Producer：生产者主界面。

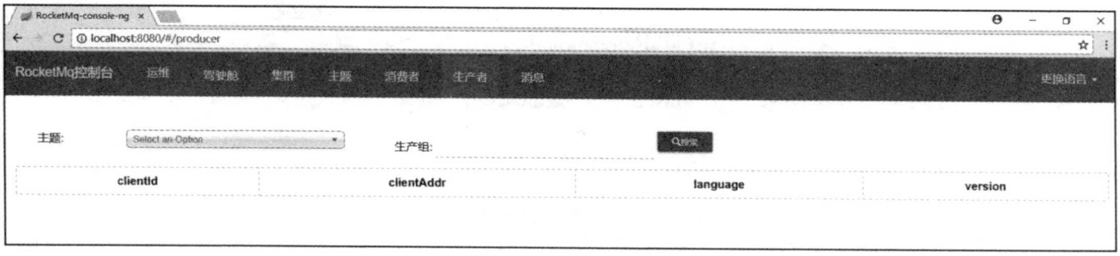

图 9-7　rocketmq-console Producer 界面

6）Message 消息查询主界面。

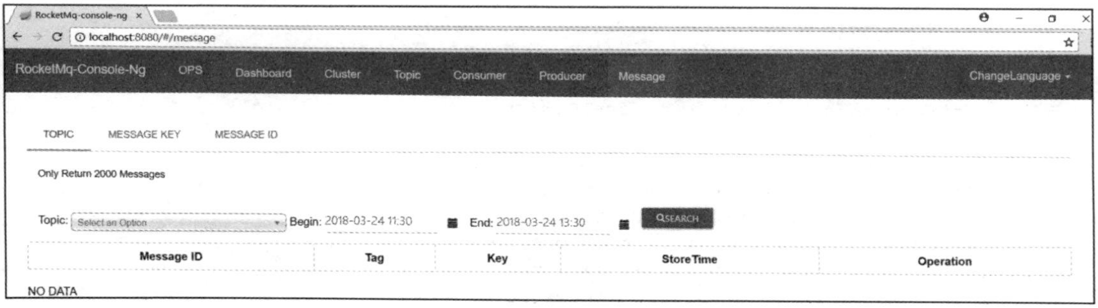

图 9-8　rocketmq-console Message 界面

rocketmq-consose 打包与部署。

修改 application.properties 与 logback.xml 中服务器端口号、NameServer 地址、rocketmq-console 监控数据目录、日志文件路径。然后执行如下命令打包：

mvn clean package -Dmaven.test.skip=true

java -jar target/rocketmq-console-ng-1.0.0.jar

rokcetmq-console 监控平台内部调用了 RocketMQ 提供的运维管理命令，下节将详细介绍。

9.7.2　RocketMQ 管理命令

在 ${ROCKETMQ_HOME}/bin/mqadmin，执行 RocketMQ 运维命令，RocketMQ 命令客户端目前支持的命令列表如下。

代码清单 9-23　MQAdminStartup#initCommand

```
public static void initCommand() {
    initCommand(new UpdateTopicSubCommand());
    initCommand(new DeleteTopicSubCommand());
    initCommand(new UpdateSubGroupSubCommand());
```

```
initCommand(new DeleteSubscriptionGroupCommand());
initCommand(new UpdateBrokerConfigSubCommand());
initCommand(new UpdateTopicPermSubCommand());
initCommand(new TopicRouteSubCommand());
initCommand(new TopicStatusSubCommand());
initCommand(new TopicClusterSubCommand());
initCommand(new BrokerStatusSubCommand());
initCommand(new QueryMsgByIdSubCommand());
initCommand(new QueryMsgByKeySubCommand());
initCommand(new QueryMsgByUniqueKeySubCommand());
initCommand(new QueryMsgByOffsetSubCommand());
initCommand(new QueryMsgByUniqueKeySubCommand());
initCommand(new PrintMessageSubCommand());
initCommand(new PrintMessageByQueueCommand());
initCommand(new SendMsgStatusCommand());
initCommand(new BrokerConsumeStatsSubCommad());
initCommand(new ConsumerConnectionSubCommand());
initCommand(new ConsumerProgressSubCommand());
initCommand(new ConsumerStatusSubCommand());
initCommand(new CloneGroupOffsetCommand());
initCommand(new ClusterListSubCommand());
initCommand(new TopicListSubCommand());
initCommand(new UpdateKvConfigCommand());
initCommand(new DeleteKvConfigCommand());
initCommand(new WipeWritePermSubCommand());
initCommand(new ResetOffsetByTimeCommand());
initCommand(new UpdateOrderConfCommand());
initCommand(new CleanExpiredCQSubCommand());
initCommand(new CleanUnusedTopicCommand());
initCommand(new StartMonitoringSubCommand());
initCommand(new StatsAllSubCommand());
initCommand(new AllocateMQSubCommand());
initCommand(new CheckMsgSendRTCommand());
initCommand(new CLusterSendMsgRTCommand());
initCommand(new GetNamesrvConfigCommand());
initCommand(new UpdateNamesrvConfigCommand());
initCommand(new GetBrokerConfigCommand());
initCommand(new QueryConsumeQueueCommand());
}
```

1. 创建或更新主题 (UpdateTopic)

1）实现类：org.apache.rocketmq.tools.command.topic.UpdateTopicSubCommand。

2）参数一览表。

表 9-1 updateTopic 命令参数一览表

参数名称	是否必填	说明
-n	是	NameServer 地址
-h	否	打印命令帮助信息
-b	-b、-c 必须一个不为空	Broker 地址，表示主题只在指定的 Broker 服务器上创建

(续)

参数名称	是否必填	说明
-c	-b、-c 必须一个不为空	Broker 集群名称，如果 -b 不为空该参数不生效。会依次从 NameServer 根据集群名称获取集群下所有 Master
-t	是	主题名称
-r	否	读队列个数，默认 4 个
-w	否	写队列个数，默认 4 个
-p	否	队列权限，默认为 6，表示可读可写
-o	否	是否是顺序消息主题，默认为 false

3）实现概述

-b：通过 Broker 地址直接定位 Broker 服务器。

-c：需要向 NameServer 发送 GET_BROKER_CLUSTER_INFO 命令获取 BrokerMaster 服务器地址，然后向 Broker 发送 UPDATE_AND_CREATE_TOPIC 命令。Broker 存储主题配置信息的默认路径为：${ROCKETMQ_HOME}/store/config/topic.json，Broker 通过与 NameServer 的心跳包将主题与 Broker 队列信息上报给 NameServer，即 Topic 的路由信息。

4）使用示例

由于是第一个命令，给出使用案例，其他类似。

```
$ sh mqadmin updateTopic -c 127.0.0.1:9876 -t PRS_PRE_BILL
```

2. 删除主题 (deleteTopic)

1）实现类：org.apache.rocketmq.tools.command.topic.DeleteTopicSubCommand。

2）参数一览表

表 9-2 deleteTopic 命令参数一览表

参数名称	是否必填	说明
-n	是	NameServer 地址
-h	否	打印命令帮助信息
-c	是	Broker 集群名称地址
-t	是	主题名称

3）实现概述

从 NameServer 获取当前所有 Broker，依次发送 DELETE_TOPIC_IN_NAMESRV，从 Broker 中删除 Topic 的配置信息，然后通过与 NameServer 的心跳机制更新 NameServer 关于主题的路由信息。

3. 创建或更新消费组配置信息 (updateSubGroup)

1）实现类：org.apache.rocketmq.tools.command.consumer.UpdateSubGroupSubCommand。

2）参数一览表。

表 9-3 updateSubGroup 命令参数一览表

参数名称	是否必填	说明
-n	是	NameServer 地址
-h	否	打印命令帮助信息
-b	-b、-c 必须一个不为空	Broker 地址，表示主题只在指定的 Broker 服务器上创建
-c	-b、-c 必须一个不为空	Broker 集群名称，如果 -b 不为空该参数不生效。会依次从 NameServer 根据集群名称获取集群下所有 Master
-g	是	消费组名称
-s	否	设置消息组是否允许消息消费。参数值为 true|false
-m	否	设置消费组是否可以从最小偏移量开始消息消费，参数值为 true|false
-d	否	设置消息消费组是否可以开启广播模式消费，参数值为 true|false
-q	否	消息消费组重试队列个数，默认为 1
-r	否	消息消费组最大重试次数，默认 16
-w	否	主服务器消息消费缓慢时由哪个从服务器承担读请求，默认为 1
-i	否	brokerId，设置该消费组默认拉取主服务器 ID，默认为 0
-a	否	当消息消费者个数发送变化后是否立即通知客户端重新进行消息队列分配，参数值为 true|false，默认为 true

3）实现概述

根据 -b、-c 定位 Broker 地址，发送 UPDATE_AND_CREATE_SUBSCRIPTIONGROUP 命令，消息消费组订阅信息在 Broker 端默认的存储路径为：${ROCKETMQ_HOME}/store/config/subscriptionGroup.json。

4. 删除消费组配置信息 (deleteSubGroup)

1）实现类：org.apache.rocketmq.tools.command.consumer.DeleteSubscriptionGroupCommand

2）参数一览表

表 9-4 deleteSubGroup 命令参数一览表

参数名称	是否必填	说明
-n	是	NameServer 地址
-h	否	打印命令帮助信息
-b	-b、-c 必须一个不为空	Broker 地址，表示主题只在指定的 Broker 服务器上创建
-c	-b、-c 必须一个不为空	Broker 集群名称，如果 -b 不为空该参数不生效。会依次从 NameServer 根据集群名称获取集群下所有 Master
-g	是	消费组名称

3）实现概述

根据 -b、-c 定位 Broker 地址，发送 DELETE_SUBSCRIPTIONGROUP。

5. 更新 Broker 配置信息 (updateBrokerConfig)

1）实现类：org.apache.rocketmq.tools.command.broker.UpdateBrokerConfigSubCommand

2）参数一览表

表 9-5　updateBrokerConfig 命令参数一览表

参数名称	是否必填	说明
-n	是	NameServer 地址
-h	否	打印命令帮助信息
-b	-b、-c 必须一个不为空	Broker 地址，表示主题只在指定的 Broker 服务器上创建
-c	-b、-c 必须一个不为空	Broker 集群名称，如果 -b 不为空该参数不生效。会依次从 NameServer 根据集群名称获取集群下所有 Master
-k	是	配置项参数名
-v	是	配置项属性值名

3）实现概述

根据 -b、-c 定位 Broker 地址，然后发送 UPDATE_BROKER_CONFIG 到 Broker 服务器，如果 -k 指定的配置项已经配置，则更新，否则忽略本次更新动作，该命令更新后的配置属性将持久化到配置文件中。

6. 更新 Broker 配置信息 (updateTopicPerm)

1）实现类：org.apache.rocketmq.tools.command.topic.UpdateTopicPermSubCommand

2）参数一览

表 9-6　updateTopicPerm 命令参数一览表

参数名称	是否必填	说明
-n	是	NameServer 地址
-h	否	打印命令帮助信息
-b	-b、-c 必须一个不为空	Broker 地址，表示主题只在指定的 Broker 服务器上创建
-c	-b、-c 必须一个不为空	Broker 集群名称，如果 -b 不为空该参数不生效。会依次从 NameServer 根据集群名称获取集群下所有 Master
-t	是	主题名称
-p	是	主题队列权限，可选值：2:W; 4:R; 6:RW

3）实现概述

根据 -b、-c 定位 Broker 地址，然后发送 UPDATE_AND_CREATE_TOPIC 命令。

7. 查看 topic 路由信息 (topicRoute)

1）实现类：org.apache.rocketmq.tools.command.topic.TopicRouteSubCommand

2）参数一览表

表 9-7　topicRoute 命令参数一览表

参数名称	是否必填	说明
-n	是	NameServer 地址
-h	否	打印命令帮助信息
-t	是	主题名称

3）实现概述

访问 NameServer 返回 topic 的路由信息。

8. 查看 Topic 消息消费队列状态 (topicStatus)

1）实现类：org.apache.rocketmq.tools.command.topic.TopicStatusSubCommand

2）参数一览表

表 9-8　topicStatus 命令参数一览表

参数名称	是否必填	说明
-n	是	NameServer 地址
-h	否	打印命令帮助信息
-t	是	主题名称

3）实现概述

首先获取主题的路由信息，然后向 Broker 发送 GET_TOPIC_STATS_INFO 获取该主题在每一个 Broker 上的配置信息并返回主题的队列信息，返回结果以消息消费组分组，返回结构如图 9-9 所示。

[BatchTest]Status			
Queue	minOffset	maxOffset	lastUpdateTimeStamp
MessageQueue [topic=BatchTest, brokerName=broker-a, queueId=3]	0	0	1970-01-01 08:00:00
MessageQueue [topic=BatchTest, brokerName=broker-a, queueId=2]	0	3	2018-03-22 21:53:26
MessageQueue [topic=BatchTest, brokerName=broker-a, queueId=1]	0	0	1970-01-01 08:00:00
MessageQueue [topic=BatchTest, brokerName=broker-a, queueId=0]	0	3	2018-03-22 21:54:49

图 9-9　获取主题状态返回结果界面

9. 获取 topic 所在 Broker 集群信息 (topicClusterList)

1）实现类：org.apache.rocketmq.tools.command.topic.TopicClusterSubCommand

2）参数一览表

表 9-9　topicClusterList 命令参数一览表

参数名称	是否必填	说明
-n	是	NameServer 地址
-h	否	打印命令帮助信息
-t	是	主题名称

3)实现概述

向 NameServer 发送 GET_BROKER_CLUSTER_INFO 获取 topic 的集群信息。

10. 获取 Broker 运行时统计信息 (brokerStatus)

1)实现类:org.apache.rocketmq.tools.command.broker.BrokerStatusSubCommand

2)参数一览表

表 9-10 brokerStatus 命令参数一览表

参数名称	是否必填	说明
-n	是	NameServer 地址
-h	否	打印命令帮助信息
-b	-b、-c 必须一个不为空	Broker 地址,表示主题只在指定的 Broker 服务器上创建
-c	-b、-c 必须一个不为空	Broker 集群名称,如果 -b 不为空该参数不生效。会依次从 NameServer 根据集群名称获取集群下所有 Master

3)实现概述

根据 -b 或 -c 定位到 Broekr 地址并发送 GET_BROKER_RUNTIME_INFO 命令获取 Broker 运行状态,使用示例:

$ sh mqadmin brokerStatus -n 127.0.0.1:9876 -c DefaultCluster

11. 根据消息 ID 查询消息 (queryMsgById)

1)实现类:org.apache.rocketmq.tools.command.message.QueryMsgByIdSubCommand

2)参数一览表

表 9-11 queryMsgById 命令参数一览表

参数名称	是否必填	说明
-n	是	NameServer 地址
-h	否	打印命令帮助信息
-i	是	消息 ID,多条消息 ID 用英文逗号分隔
-g	否	消息消费组名称
-d	否	消息消费者 ID。-g -d 参数不同时为空时生效,此时 -s 参数失效,其主要作用是指定消费组内指定的消息消费者消费该批消息
-s	否	如果为 true 表示再次发送该消息

3)实现概述

该命令集合了消息发送、消息查询、消息消费三个功能。

-i 消息 ID,在没有其他参数的情况下,表示查询该批消息。

-i -g -d:消息消费。

-s:如果设置为 true,重新将该消息发送到 Broker 服务器。

12. 根据消息索引键查询消息 (queryMsgByKey)

1)实现类:org.apache.rocketmq.tools.command.message.QueryMsgByKeySubCommand

2）参数一览表

表 9-12　queryMsgByKey 命令参数一览表

参数名称	是否必填	说明
-n	是	NameServer 地址
-h	否	打印命令帮助信息
-t	是	消息所属主题
-k	否	消息索引键，只支持单个

3）实现概述

从 NameServer 获取主题的路由信息，然后并发向 Broker 发送 QUERY_MESSAGE 查询消息，待该 Topic 所有 Broker 返回结果合并后返回。

13. 根据消息唯一键查询消息 (queryMsgByUniqueKey)

1）实现类：org.apache.rocketmq.tools.command.message.QueryMsgByUniqueKeySubCommand

2）参数一览表

表 9-13　queryMsgByUniqueKey 命令参数一览表

参数名称	是否必填	说明
-n	是	NameServer 地址
-h	否	打印命令帮助信息
-i	是	消息唯一键，多条消息 ID 用英文逗号分隔
-g	否	消息消费组名称
-d	否	消息消费者 ID。-g -d 参数不同时为空时生效。直接使用指定的消息消费者消费该批消息
-t	是	消息主题

3）实现概述

该命令根据主题与消息唯一键查询消息，如果 -d -g 不为空时，将直接消费该条消息。

14. 打印消息 (printMsg)

1）实现类：org.apache.rocketmq.tools.command.message.PrintMessageSubCommand

2）参数一览表

表 9-14　printMsg 命令参数一览表

参数名称	是否必填	说明
-n	是	NameServer 地址
-h	否	打印命令帮助信息
-t	是	消息主题
-c	否	字符编码，可选值：UTF-8,GBK
-s	否	消息过滤表达式
-b	否	开始时间戳，支持 long 或 yyyy-MM-dd#HH:mm:ss:SSS
-e	否	结束时间戳，支持 long 或 yyyy-MM-dd#HH:mm:ss:SSS
-d	否	是否打印消息 body，可选值为 true|false

3）实现概述

本命令其实是实现基于时间戳的消息查询，根据主题的路由信息获取该主题的消息消费队列，循环处理每一个消息消费队列。如果 -b，-e 为空，则从消息消费队列的最小偏移量开始，否则根据 -b，-e 时间戳查询出最小偏移量、最大偏移量，从每个消息消费每次最多拉取 32 条消息并打印输出。

15. 根据消息队列打印消息 (printMsgByQueue)

1）实现类：org.apache.rocketmq.tools.command.message.PrintMessageByQueueCommand

2）参数一览表

表 9-15　printMsgByQueue 命令参数一览表

参数名称	是否必填	说明
-n	是	NameServer 地址
-h	否	打印命令帮助信息
-t	是	消息主题
-a	是	Broker 名称
-i	是	消息队列 ID
-c	否	字符编码，可选值：UTF-8,GBK
-s	否	消息过滤表达式
-b	否	开始时间戳，支持 long 或 yyyy-MM-dd#HH:mm:ss:SSS
-e	否	结束时间戳，支持 long 或 yyyy-MM-dd#HH:mm:ss:SSS
-p	否	是否打印消息，可选值 true|false，默认为 false
-d	否	是否打印消息 body，可选值为 true|false
-f	否	是否打印 tag 信息，可选值为 true|false，默认为 fasle

3）实现概述

根据主题、Broker 名称、消息消费队列构建 MessageQueue 对象，从消息服务器根据拉取偏移量循环从服务器拉取消息。

16. 测试 Broker 消息发送性能 (sendMsgStatus)

1）实现类：org.apache.rocketmq.tools.command.broker.SendMsgStatusCommand

2）参数一览表

表 9-16　sendMsgStatus 命令参数一览表

参数名称	是否必填	说明
-n	是	NameServer 地址
-b	是	Broker 名称
-s	否	每次消息发送字节数，默认 128
-c	否	默认运行 50 次后结束

3）实现概述

模拟消息发送，向主题名为 Broker 名称的主题发送消息，测试消息发送的时间。
示例：`$ sh mqadmin sendMsgStatus -n 127.0.0.1:9876 -b broker-a -c 3`

代码清单 9-24　sendMsgStatus 输出示例

```
rt:7ms, SendResult=SendResult [sendStatus=SEND_OK,
msgId=C0A801033AA818B4AAC27AAC1C4E0001,
offsetMsgId=C0A8010300002A9F00000000000005C9, messageQueue=MessageQueue
[topic=broker-a, brokerName=broker-a, queueId=0], queueOffset=1]
rt:14ms, SendResult=SendResult [sendStatus=SEND_OK,
msgId=C0A801033AA818B4AAC27AAC1C550002,
offsetMsgId=C0A8010300002A9F00000000000006DA, messageQueue=MessageQueue
[topic=broker-a, brokerName=broker-a, queueId=0], queueOffset=2]
rt:5ms, SendResult=SendResult [sendStatus=SEND_OK,
msgId=C0A801033AA818B4AAC27AAC1C630003,
offsetMsgId=C0A8010300002A9F00000000000007EB, messageQueue=MessageQueue
[topic=broker-a, brokerName=broker-a, queueId=0], queueOffset=3]
```

17. 查看 Broker 消费状态 (brokerConsumeStats)

1）实现类：org.apache.rocketmq.tools.command.broker.BrokerConsumeStatsSubCommad

2）参数一览表

表 9-17　brokerConsumeStats 命令参数一览表

参数名称	是否必填	说明
-n	是	NameServer 地址
-b	是	Broker 地址
-t	否	请求超时时间，默认为一直等待，单位：毫秒
-l	否	延迟条数，如果设置该值，如果未消费消息条数小于该值则不显示
-o	否	是否是顺序消息，可选值为 true|false，默认为 false

3）实现概述

根据 Broker 上的订阅消息组反推出所有消息组订阅的主题，然后统计各消费组在该 Broker 上消息消费队列的消息消费进度，命令执行结果如图 9-10 所示。

```
#Topic      #Group              #Broker Name  #QID  #Broker Offset  #Consumer Offset  #Diff  #LastTime
TopicTest   PRS_BILL_CONSUME    broker-a      0     1               1                 0      2018-03-22 20:54:29
Diff Total: 0
```

图 9-10　brokerConsumeStats 运行结果图

Topic：主题名称。
Group：消息消费组。
Broker Name：Broker 名称。
Broker Offset：Broker 消息消费队列当前偏移量。
Consume Offset：该消息消费组当前消息消费进度。

Diff：Brokeroffset-ConsumeOffset，消息滞留条数。

如果 -1 3 则表示如果 BrokerOffset-ConsmeOffset 小于 1 的条目不显示。

18. 查看消费组连接信息 [消息消费者](consumerConnection)

1）实现类：org.apache.rocketmq.tools.command.connection.ConsumerConnectionSubCommand

2）参数一览表

表 9-18　consumerConnection 命令参数一览表

参数名称	是否必填	说明
-n	是	NameServer 地址
-g	是	消息消费组名称

3）实现概述

构建消息消费组的重试主题（%RETRY%+ 消息消费组名），从 NameServer 获取该主题的路由信息，从中选择一个 Broker，返回消息消费组内所有与该 Broker 建立的长连接信息，返回结果如图 9-11 所示。

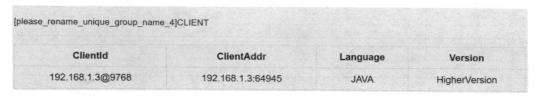

图 9-11　consumerConnection 运行结果图

19. 查看消费组处理进度 [消息消费进度](consumerProgress)

1）实现类：org.apache.rocketmq.tools.command.consumer.ConsumerProgressSubCommand

2）参数一览表

表 9-19　consumerProgress 命令参数一览表

参数名称	是否必填	说明
-n	是	NameServer 地址
-g	否	消息消费组名称

3）实现概述

-g 查看指定消息消费组的消息消费进度，如果不指定 -g 选项，则从 NameServer 获取所有的主题，并从中筛选出所有重试主题，然后从重试主题中采用提取字符串的方式提取到消费组，然后根据消息消费组查看该组的消息消费进度。

20. 查看消息消费组内部线程状态 (consumerStatus)

1）实现类：org.apache.rocketmq.tools.command.consumer.ConsumerStatusSubCommand

2）参数一览表

表 9-20　consumerStatus 命令参数一览表

参数名称	是否必填	说明
-n	是	NameServer 地址
-g	是	消息消费组名称
-i	否	消息消费者 ID
-s	否	是否打印 jstack 命令

3）实现概述

获取消息消费组内所有的消息消费客户端连接信息，可以通过 -i 只显示指定的消息消费者，-s 可以指定是否打印 jstack 线程栈，其实现是调用 Thread.getAllStackTraces() 方法。

21. 克隆消费组进度 (cloneGroupOffset)

1）实现类：org.apache.rocketmq.tools.command.offset.CloneGroupOffsetCommand

2）参数一览表

表 9-21　cloneGroupOffset 命令参数一览表

参数名称	是否必填	说明
-n	是	NameServer 地址
-s	是	原消息消费组
-d	是	目的消息消费组
-t	是	主题名称

3）实现概述

将源消息消费组消费进度复制到目的消息组进度，就是用源消息消费组的消息消费进度更新目的消息消费组的消息进度。

22. 查看所有集群下 Broker 运行状态 (clusterList)

1）实现类：org.apache.rocketmq.tools.command.cluster.ClusterListSubCommand

2）参数一览表

表 9-22　clusterList 命令参数一览表

参数名称	是否必填	说明
-n	是	NameServer 地址
-m	否	是否打印 Broker 所有运行状态，默认为 false
-i	否	循环打印间隔，默认为秒

3）实现概述

该命令是 brokerStatus 的全量信息，其运行如图 9-12 所示。

23. 查看所有主题信息 (topicList)

1）实现类：org.apache.rocketmq.tools.command.topic.TopicListSubCommand

图 9-12 clusterList 运行结果图

2）参数一览表

表 9-23 topicList 命令参数一览表

参数名称	是否必填	说明
-n	是	NameServer 地址
-c	否	Broker 集群名称

3）实现概述

从 NameServer 获取所有主题列表。-c 表示只返回该 Broker 集群上存在的主题。与主题相关的命令如图 9-13 所示。

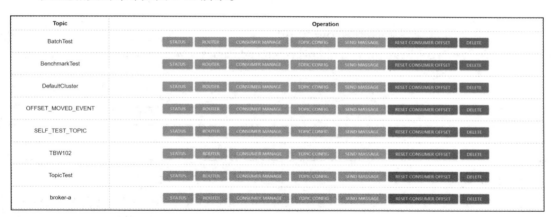

图 9-13 topicList 相关功能图

24. 更新 NameServer KV 配置 (updateKvConfig)

1）实现类：org.apache.rocketmq.tools.command.namesrv.UpdateKvConfigCommand

2）参数一览表

表 9-24 updateKvConfig 命令参数一览表

参数名称	是否必填	说明
-n	是	NameServer 地址
-s	是	配置命名空间，例如 ORDER_TOPIC_CONFIG，表示顺序主题配置信息
-k	是	配置 key
-v	是	配置 value

3)实现概述

直接将 KV 信息发送命令 PUT_KV_CONFIG 到 NameServer,更新 KV 配置项并默认持久化到 ${user.home}/namesrv/kvconfig.json,可通过在 NameServer 配置文件中设置 kvConfigPath 来改变其路径。

25. 更新 NameServer KV 配置 (deleteKvConfig)

1)实现类:org.apache.rocketmq.tools.command.namesrv.DeleteKvConfigCommand

2)参数一览表

表 9-25 deleteKvConfig 命令参数一览表

参数名称	是否必填	说明
-n	是	NameServer 地址
-s	是	配置命名空间,例如 ORDER_TOPIC_CONFIG,表示顺序主题配置信息
-k	是	配置 key

3)实现概述

向 NameServer 发送 DELETE_KV_CONFIG 命令,删除配置命名空间下键为 key 的配置项。

26. 擦除 Broker 写权限 (wipeWritePerm)

1)实现类:org.apache.rocketmq.tools.command.namesrv.WipeWritePermSubCommand

2)参数一览表

表 9-26 wipeWritePerm 命令参数一览表

参数名称	是否必填	说明
-n	是	NameServer 地址
-b	是	Broker 名称

3)实现概述

向 NameServer 发送 WIPE_WRITE_PERM_OF_BROKER 命令,将存储在 NameServer 上的所有关于主题中分布在该 Broker 上的所有队列的权限设置为非可写,也就是拒绝向该 Broker 上写消息。

27. 重置消息消费组消费进度 [根据时间](resetOffsetByTime)

1)实现类:org.apache.rocketmq.tools.command.offset.ResetOffsetByTimeCommand

2)参数一览表

表 9-27 resetOffsetByTime 命令参数一览表

参数名称	是否必填	说明
-n	是	NameServer 地址
-g	是	消息消费组名称
-t	是	主题名称

(续)

参数名称	是否必填	说明
-s	是	时间戳，可选值：now、时间戳（毫秒）、yyyy-MM-dd#HH:mm:ss:SSS
-f	否	是否强制刷新
-c	否	C++ client offset

3）实现概述

根据主题的路由信息找出所有 Master Broker 地址，逐一发送 INVOKE_BROKER_TO_RESET_OFFSET 命令，在 Broker 端找出该主题所对应的消息消费队列，然后根据时间戳从 ConsumeQueue 中找到合适的偏移量，用该偏移量更新指定消息消费组的消息消费进度。

28. 创建、更新、删除顺序消息 KV 配置 (updateOrderConf)

1）实现类：org.apache.rocketmq.tools.command.topic.UpdateOrderConfCommand

2）参数一览表

表 9-28　updateOrderConf 命令参数一览表

参数名称	是否必填	说明
-n	是	NameServer 地址
-t	是	主题名称
-v	否，创建、更新时必填	顺序消息 KV 配置字符串，格式为 brokerName1:num;brokerName2:num
-m	否	操作方式：put：创建或更新；get：获取配置；delete：删除，默认为打印该命令的帮助信息

3）实现概述

向 NameServer 发送 PUT_KV_CONFIG 或 DELETE_KV_CONFIG 或 GET_KV_CONFIG 命令更新、删除或查询该主题的配置信息（顺序消息）。

29. 删除过期消息消费队列文件 (cleanExpiredCQ)

1）实现类：org.apache.rocketmq.tools.command.broker.CleanExpiredCQSubCommand

2）参数一览表

表 9-29　cleanExpiredCQ 命令参数一览表

参数名称	是否必填	说明
-n	是	NameServer 地址
-b	否	broker 地址
-c	否	Broker 集群名称

3）实现概述

根据 -b 或 -c 参数首先定位本次需要处理的 Broker，然后发送 CLEAN_EXPIRED_CONSUMEQUEUE 命令触发一次消息消费队列过期文件清除。

30. 删除未使用 topic(cleanUnusedTopic)

1）实现类：org.apache.rocketmq.tools.command.broker.CleanUnusedTopicCommand

2）参数一览表

表 9-30　cleanUnusedTopic 命令参数一览表

参数名称	是否必填	说明
-n	是	NameServer 地址
-b	否	broker 地址
-c	否	Broker 集群名称

3）实现概述

首先根据 -b、-c 参数定位到本次需要处理的 Broker，然后发送 CLEAN_UNUSED_TOPIC 命令，Broker 端判断主题未被使用的依据是没有消息消费者订阅该主题，此时删除该主题下的所有消息消费队列。

31. 开启 RocketMQ 监控 (startMonitoring)

1）实现类：org.apache.rocketmq.tools.command.consumer.StartMonitoringSubCommand

2）参数一览表

表 9-31　startMonitoring 命令参数一览表

参数名称	是否必填	说明
-n	是	NameServer 地址

3）实现概述

创建一个消息消费者订阅 OFFSET_MOVED_EVENT 主题，该主题下的消息为被删除的消息当该主题下有消息到达后报告消息的信息，同时会开启一个定时调度任务，重点关注消息重试的信息。具体实现类：org.apache.rocketmq.tools.monitor.MonitorService，监控时间如图 9-14 所示。

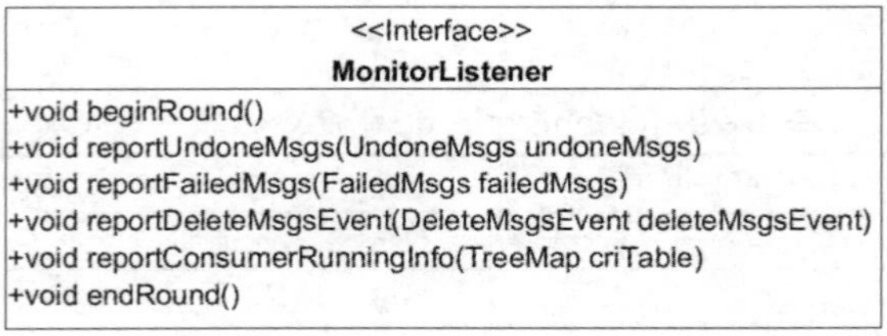

```
<<Interface>>
MonitorListener
+void beginRound()
+void reportUndoneMsgs(UndoneMsgs undoneMsgs)
+void reportFailedMsgs(FailedMsgs failedMsgs)
+void reportDeleteMsgsEvent(DeleteMsgsEvent deleteMsgsEvent)
+void reportConsumerRunningInfo(TreeMap criTable)
+void endRound()
```

图 9-14　MonitorListener 监控图

void beginRound() 一轮监控开始被调用。

void reportUndoneMsgs（UndoneMsgs undoneMsgs）：消息消费重试时调用。

void reportFailedMsgs（FailedMsgs failedMsgs）：消息发送失败时调用，默认为空实现。

void reportDeleteMsgsEvent（DeleteMsgsEvent deleteMsgsEvent）：消息被删除时调用。

void reportConsumerRunningInfo（TreeMap<String/* clientId */, ConsumerRunningInfo> criTable）：报告消息消费者运行状态，消息消费重试时被调用。

void endRound()：一轮监控结束时被调用。

32. 打印主题与消费组的 TPS 统计信息 (statsAll)

1）实现类：org.apache.rocketmq.tools.command.stats.StatsAllSubCommand

2）参数一览表

表 9-32　statsAll 命令参数一览表

参数名称	是否必填	说明
-n	是	NameServer 地址
-a	否	当前激活的主题
-t	否	指定主题名称

3）实现概述

统计 Topic 各消费组的消费 TPSX 信息。效果图如图 9-15 所示。

```
#Topic         #Consumer Group    #Accumulation    #InTPS      #OutTPS     #InMsg24Hour    #OutMsg24Hour
TopicTest      PRS BILL CONSUME   0                0.00        0.00        0               0
```

图 9-15　statsAll 监控图

Topic：主题名称。

Consume Group：消息消费组。

Accumulation：总的消息消费堆积条数。

InTPS：Broker 每分钟发送消息到该 Broker 的 TPS。

OutTPS：Broker 每分钟消息消费组消息拉取 TPS。

InMsgg24Hour：Broker 一天消息发送总数量。

OutMsg24Houre：Broker 一天总拉取条数。

33. 查看消息队列负载情况 (allocateMQ)

1）实现类：org.apache.rocketmq.tools.command.topic.AllocateMQSubCommand

2）参数一览表

表 9-33　allocateMQ 命令参数一览表

参数名称	是否必填	说明
-n	是	NameServer 地址
-t	是	主题名称
-i	是	消息消费者列表

3）实现概述

该命令主要是根据 topic 的路由信息与消息消费者列表，输出各个消息消费者队列的分配情况，分配算法使用平均分配。

34. 检测消息发送响应时间 (checkMsgSendRT)

1）实现类：org.apache.rocketmq.tools.command.message.CheckMsgSendRTCommand

2）参数一览表

表 9-34 checkMsgSendRT 命令参数一览表

参数名称	是否必填	说明
-n	是	NameServer 地址
-t	是	主题名称
-a	否	运行此时，默认 100 次
-s	否	测试消息长度默认 128 个字节

3）实现概述

构建测试消息，然后发送该主题的消息，返回该主题下所有消息消费队列的响应时间。

35. 测试所有集群消息发送响应时间 (clusterRT)

1）实现类：org.apache.rocketmq.tools.command.cluster.CLusterSendMsgRTCommand

2）参数一览表

表 9-35 clusterRT 命令参数一览表

参数名称	是否必填	说明
-n	是	NameServer 地址
-a	否	运行次数，默认为 100 次
-s	否	发送测试消息长度，默认 128 个字节
-c	否	集群名称，默认打印全部集群
-p	否	是否打印日志，默认为 false
-m	否	机器名
-i	否	运行间隔时间，默认为 10s

3）实现概述

与 checkMsgSendRT 实现类似，只是这个是发往所有的 Broker 或是指定集群名称内的所有 Broker。

36. 获取 NameServer 配置 (getNamesrvConfig)

1）实现类：org.apache.rocketmq.tools.command.namesrv.GetNamesrvConfigCommand

2）参数一览表

表 9-36 getNamesrvConfig 命令参数一览表

参数名称	是否必填	说明
-n	是	NameServer 地址

3）实现概述

向所有 NameServer 发送 GET_NAMESRV_CONFIG 命令，服务端返回 NameServer 的配置。

37. 更新 NameServer 配置 (updateNamesrvConfig)

1）实现类：org.apache.rocketmq.tools.command.namesrv.UpdateNamesrvConfigCommand

2）参数一览表

表 9-37　updateNamesrvConfig 命令参数一览表

参数名称	是否必填	说明
-n	是	NameServer 地址
-k	是	配置参数名称
-v	是	配置参数值

3）实现概述

向 NameServer 发送 UPDATE_NAMESRV_CONFIG 命令，NameServer 收到请求后将指定的参数更新 ServerConfig 运行时实例，并持久化到 NameServer 配置文件中。

38. 获取 Broker 配置信息 (getBrokerConfig)

1）实现类：org.apache.rocketmq.tools.command.broker.GetBrokerConfigCommand

2）参数一览表

表 9-38　getBrokerConfig 命令参数一览表

参数名称	是否必填	说明
-n	是	NameServer 地址
-b	-b、-c 不能同时为空	Broker 地址
-c	-b、-c 不能同时为空	集群名称

3）实现概述

首先根据 -b、-c 参数定位 Broker 地址，然后发送 GET_BROKER_CONFIG 命令获取 Broker 配置属性。

39. 查询消息消费队列 (queryCq)

1）实现类：org.apache.rocketmq.tools.command.queue.QueryConsumeQueueCommand

2）参数一览表

表 9-39　queryCq 命令参数一览表

参数名称	是否必填	说明
-n	是	NameServer 地址
-t	是	主题名称
-q	是	队列 ID，从 0 开始
-i	是	从消息队列开始位置

参数名称	是否必填	说明
-c	否	消息队列最大偏移量，默认为 10
-b	否	Broker 地址
-g	否	消息消费组名称

3）实现概述

如果没有指定 -b 选项，则选择该 Topic 路由信息表中第一个 Broker 地址，发送 QUERY_CONSUME_QUEUE 命令，Broker 返回该消息消费队列的最大、最小偏移量；如果消息消费组不为空，则返回该消息消费组的订阅信息；如果订阅信息不为空，则返回过滤器信息；然后返回查询到的消息消费条目（物理偏移量、消息长度、TAG HashCode）。

9.8 应用场景分析

随着互联网技术蓬勃发展、微服务架构思想的兴起，系统架构追求小型化、轻量化，原有的大型集中式的 IT 系统通常需要进行垂直拆分，孵化出颗粒度更小的众多小型系统，因此对系统间松耦合的要求越来越高，目前 RPC、服务治理、消息中间件几乎成为互联网架构的标配。

引入消息中间件，服务之间可以通过可靠的异步调用，从而降低系统之间的耦合度，提高系统的可用性。消息中间件另一个重要的应用场景是解决系统之间数据的一致性（最终一致性）。

消息中间件的两大核心点：异步与解耦。

经典应用场景：数据同步。

某公司的 IT 系统由多个子系统构成，例如基础数据平台、社交平台、订单平台。各个子系统之间使用单独的数据库，这样就面临基础数据的管理，比如用户信息表、行政区域表，按照职责划分的话这些数据是在基础数据平台进行维护的，但各个子系统的业务表需要关联这些基础数据，所有基础数据的表结构也会同时存在各个子系统之间，此时数据的同步如何解决呢？

解决方案 1：基础平台数据发送变化后，通过调用社交平台、订单平台提供的数据同步接口完成数据同步。

缺点：基础平台必须依赖各个业务子系统，当业务系统增加时，需要修改基础平台同步代码或配置文件，基础平台与各个业务子系统强耦合。

解决方案 2：引入消息中间件，基础平台在数据发送变化后，发送一条消息到消息服务器，然后就正常返回。

各个业务系统订阅数据同步消息主题，自己负责消息消费，并依靠 MQ 提供的消息重试机制方便异常处理。

9.9 本章小结

本章从如何使用 RocketMQ 入手，介绍了 RocketMQ 批量消息发送、发送消息队列选择、消息过滤（基于 TAG、SQL92、类过滤模式）的使用、RocketMQ 事务消息，并介绍了 RocketMQ 与 Spring、Spring Cloud 的整合，然后详细讲解了 RocketMQ 监控平台的搭建并且详细介绍了 RocketMQ 当前支持的 39 个监控命令。

附录 A 参数说明

表 A-1 NameServer 配置属性

参数名	参数类型	描述
rocketmqHome	String	RocketMQ 主目录,默认为用户主目录
namesrvAddr	String	NameServer 地址
kvConfigPath	String	kv 配置文件路径,包含顺序消息主题的配置信息
configStorePath	String	NameServer 配置文件路径,建议使用 -c 指定 NameServer 配置文件路径
clusterTest	boolean	是否开启集群测试,默认为 false
orderMessageEnable	boolean	是否支持顺序消息,默认为 false

表 A-2 NameServer、Broker 网络配置属性

参数名	参数类型	描述
listenPort	int	服务端监听端口
serverWorkerThreads	int	Netty 业务线程池线程个数
serverCallbackExecutorThreads	int	Netty public 任务线程池线程个数,Netty 网络设计,根据业务类型会创建不同的线程池,比如处理发送消息、消息消费、心跳检测等。如果该业务类型(RequestCode)未注册线程池,则由 public 线程池执行
serverSelectorThreads	int	IO 线程池线程个数,主要是 NameServer、Broker 端解析请求、返回响应的线程个数,这类线程主要是处理网络请求的,解析请求包,然后转发到各个业务线程池完成具体的业务操作,然后将结果再返回调用方
serverOnewaySemaphoreValue	int	send oneway 消息请求并发度
serverAsyncSemaphoreValue	int	异步消息发送最大并发度

（续）

参数名	参数类型	描述
serverChannelMaxIdleTimeSeconds	int	网络连接最大空闲时间，默认 120s，如果连接空闲时间超过该参数设置的值，连接将被关闭
serverSocketSndBufSize	int	网络 socket 发送缓存区大小，默认 64K
serverSocketRcvBufSize	int	网络 socket 接收缓存区大小，默认 64K
serverPooledByteBufAllocatorEnable	boolean	ByteBuffer 是否开启缓存，建议开启
useEpollNativeSelector	boolean	是否启用 Epoll IO 模型，Linux 环境建议开启

表 A-3　Broker 配置属性（服务器属性）

参数名	参数类型	描述
rocketmqHome	String	RocketMQ 主目录，默认为用户主目录
namesrvAddr	String	nameServer 地址
brokerIP1	String	Broker 服务地址
brokerIP2	String	Broker HA IP 地址，供 slave 同步消息的地址
brokerName	String	Broker 服务器名称，默认为服务器 hostname
brokerClusterName	String	Broker 集群名称，默认 DefaultCluster
brokerId	int	BrokerID,0 表示主节点，大于 0 表示从节点
brokerPermission	Int	Broker 权限，默认为 6，表示可读可写
defaultTopicQueueNums	int	主题在一个 Broker 上创建队列数量，默认为 8 个
autoCreateTopicEnable	boolean	是否自动创建主题，默认为 true
clusterTopicEnable	boolean	集群名称是否可用在主题使用，默认为 true
brokerTopicEnable	boolean	Broker 名称是否可以用作主题使用，默认为 true
autoCreateSubscriptionGroup	boolean	是否自动创建消费组订阅配置信息
messageStorePlugIn	String	消息存储插件地址，默认为空字符串
sendMessageThreadPoolNums	int	服务端处理消息发送线程池线程数量。默认为 1
pullMessageThreadPoolNums	int	服务端处理消息拉取线程池线程数量，默认为 16 加上当前操作系统 CPU 核数的两倍
queryMessageThreadPoolNums	int	服务端处理查询消息线程池线程数量，默认为 8 加上当前操作系统 CPU 核数
adminBrokerThreadPoolNums	int	服务端处理控制台管理命令线程池线程数量，默认为 16
clientManageThreadPoolNums	int	服务端处理客户端管理（心跳、注册、取消注册）线程池线程线程数量，默认 32
consumerManageThreadPoolNums	int	服务端处理消费管理（获取消费者列表、更新消费进度、查询消费进度等）线程池线程数量，默认 32
flushConsumerOffsetInterval	int	持久化消息消费进度（consumerOffset.json）文件的频率，默认 5s
flushConsumerOffsetHistoryInterval	int	当前版本未使用
rejectTransactionMessage	boolean	是否拒绝事务消息，默认为 false
fetchNamesrvAddrByAddressServer	boolean	是否支持从服务器获取 NameServer 地址
sendThreadPoolQueueCapacity	int	消息发送线程池任务队列初始大小，默认为 10000

（续）

参数名	参数类型	描述
pullThreadPoolQueueCapacity	int	消息拉取线程池任务队列初始大小，默认 100000
queryThreadPoolQueueCapacity	int	查询消息线程池任务队列初始大小，默认 20000
clientManagerThreadPoolQueueCapacity	int	客户端管理线程池任务队列初始大小，默认 1000000
consumerManagerThreadPoolQueueCapacity	int	消费管理线程池任务队列初始大小，默认 1000000
filterServerNums	int	Broker 服务器过滤服务器数量，默认为 0
longPollingEnable	boolean	是否开启长轮询，默认 true
shortPollingTimeMills	long	短轮询等待时间，默认 1s
notifyConsumerIdsChangedEnable	boolean	消费者数量变化后是否立即通知 RebalcneService 线程，以便进行消息队列重新负载。默认为 true
highSpeedMode	boolean	当前版本未使用
transferMsgByHeap	boolean	消息传输是否使用堆内存，默认为 true
maxDelayTime	int	当前版本未使用
regionId	String	消息区域，默认为 DefaultRegion
registerBrokerTimeoutMills	int	注册 Broker 超时时间，默认为 6s
slaveReadEnable	boolean	从节点是否可读，默认为 false
disableConsumeIfConsumerReadSlowly	boolean	如果消费组消息消费堆积是否禁用该消费组继续消费消息，默认为 false
consumerFallbehindThreshold	long	消息消费堆积阈值，默认为 16G，在 disableConsumeIfConsumerReadSlowly 为 true 时生效
brokerFastFailureEnable	boolean	是否支持 Broker 快速失败，如果为 true 表示会立即清除发送消息线程池、消息拉取线程池中排队任务，直接返回系统错误，默认为 true
waitTimeMillsInSendQueue	long	清除发送线程池任务队列的等待时间，如果系统时间减去任务放入队列中的时间小于 waitTimeMillsInSendQueue，本次请求任务暂不移除该任务。默认为 200ms
waitTimeMillsInPullQueue	long	清除消息拉取线程池任务队列的等待时间，如果系统时间减去任务放入队列中的时间小于 waitTimeMillsInPullQueue，本次请求任务暂不移除该任务。默认为 5s
filterDataCleanTimeSpan	long	清除过滤数据的时间间隔，默认为 24h
filterSupportRetry	boolean	消息过滤是否支持重试，默认为 false
enablePropertyFilter	boolean	是否支持根据属性过滤，默认为 false，如果使用基于标式 SQL92 模式过滤消息，则该参数必须设置为 true

表 A-4 Broker 配置属性（存储相关属性）

参数名	参数类型	描述
storePathRootDir	String	broker 存储目录，默认为用户的主目录 /store，建议配置
storePathCommitLog	String	Commitlog 存储目录，默认为 ${ storePathRootDir }/commitlog

(续)

参数名	参数类型	描述
mapedFileSizeCommitLog	int	单个 commitlog 文件的大小，默认为 1G
mapedFileSizeConsumeQueue	int	单个 consumequeue 文件的大小，默认为 30W * 20（字节）。表示单个 ConsumeQueue 文件中存储 30W 个 ConsumeQueue 条目
enableConsumeQueueExt	boolean	是否启用 ConsumeQueue 扩展属性，默认为 false
mappedFileSizeConsumeQueueExt	int	ConsumeQueue 扩展文件大小，默认为 48M
bitMapLengthConsumeQueueExt	int	ConsumeQueue 扩展过滤 bitmap 大小，默认为 64
flushIntervalCommitLog	int	commitlog 刷盘频率，默认 500ms
commitIntervalCommitLog	int	commitlog 提交频率，默认 200ms
useReentrantLockWhenPutMessage	boolean	消息存储到 commitlog 文件时获取锁类型，如果为 true，使用 ReentrantLock，否则使用自旋锁。默认为 false
flushCommitLogTimed	boolean	默认为 false, 表示 await 方法等待 flushIntervalCommitLog，如果为 true, 表示使用 Thread.sleep 方法等待
flushIntervalConsumeQueue	int	consumuQueue 文件刷盘频率，默认为 1s
cleanResourceInterval	int	清除过期文件线程调度频率，默认每隔 10s 检测一下是否需要清除过期文件
deleteCommitLogFilesInterval	int	删除 Commitlog 文件的间隔时间，删除一个文件后，等一下再删除下一个文件。默认为 100ms
deleteConsumeQueueFilesInterval	int	删除 ConsumeQueue 文件的时间间隔，默认为 100ms
destroyMapedFileIntervalForcibly	int	销毁 MappedFile 被拒绝的最大从存活时间，默认为 120s。清除过期文件线程在初次销毁 MappedFile 时，如果该文件被其他线程 (引用次数大于 0)，则设置 MappedFile 的可用状态为 false, 并设置第一次删除时间，下一次清理任务到达时，如果系统时间大于初次删除时间加上 destroyMapedFileIntervalForcibly，则将 ref 次数一次减 1000，直到引用次数小于 0，则释放物理资源
redeleteHangedFileInterval	int	重试删除文件间隔，默认为 120s，配合 destroyMapedFileIntervalForcibly 使用
deleteWhen	String	磁盘文件充足的情况下，默认每天的什么时候执行删除过期文件。默认为 04，表示凌晨 4 点
diskMaxUsedSpaceRatio	int	commitlog 目录所在分区的最大使用比例，如果 commitlog 目录所在的分区使用比例大于该值，则触发过期文件删除。默认 75
fileReservedTime	int	文件保留时间，默认 72 小时，表示非当前写文件最后一次更新时间加上 fileReservedTime 小于当前时间，该文件将被清理
putMsgIndexHightWater	int	当前版本未使用
maxMessageSize	int	默认允许的最大消息体，默认为 4M
checkCRCOnRecover	boolean	文件恢复时是否校验 CRC，默认为 true
flushCommitLogLeastPages	int	一次刷盘至少需要脏页的数量，默认 4 页，针对 CommitLog 文件

(续)

参数名	参数类型	描述
commitCommitLogLeastPages	int	一次提交至少需要脏页的数量，默认 4 页，针对 CommitLog 文件
flushLeastPagesWhenWarmMapedFile	int	用字节 0 填充整个文件的，每多少页刷盘一次。默认 4096 页，异步刷盘模式生效
flushConsumeQueueLeastPages	int	一次刷盘至少需要脏页的数量，默认 2 页，针对 Consume 文件
flushCommitLogThoroughInterval	int	commitlog 两次刷盘的最大间隔，如果超过该间隔，将忽略 flushCommitLogLeastPages 要求直接执行刷盘操作，默认 10s
commitCommitLogThoroughInterval	int	Commitlog 两次提交的最大间隔，如果超过该间隔，将忽略 commitCommitLogLeastPages 直接提交。默认 200ms
flushConsumeQueueThoroughInterval	int	Consume 两次刷盘的最大间隔，如果超过该间隔，将忽略 flushConsumeQueueLeastPages 直接刷盘，默认 60s
maxTransferBytesOnMessageInMemory	int	一次服务端消息拉取，消息在内存中传输允许的最大传输字节，默认为 256K
maxTransferCountOnMessageInMemory	int	一次服务消息拉取，消息在内存中传输运行的最大消息调试，默认 32 条
maxTransferBytesOnMessageInDisk	int	一次服务消息端消息拉取，消息在磁盘中传输允许的最大字节，默认 64K
maxTransferCountOnMessageInDisk	int	一次消息服务端消息拉取，消息在磁盘中传输允许的最大条数，默认为 8 条
accessMessageInMemoryMaxRatio	int	访问消息在内存中比率，默认为 40
messageIndexEnable	boolean	是否支持消息索引文件，默认为 true
maxHashSlotNum	int	单个索引文件 hash 槽的个数，默认为五百万
maxIndexNum	int	单个索引文件索引条目的个数，默认为两千万
maxMsgsNumBatch	int	一次查询消息最大返回消息条数，默认 64 条
messageIndexSafe	boolean	消息索引是否安全，默认为 false，文件恢复时选择文件检测点 (commitlog,consumeque) 的最小值与文件最后更新对比，如果为 true，文件恢复时选择文件检测点保存的索引更新时间作为对比
haListenPort	int	Master 监听端口，从服务器连接该端口，默认为 10912
haSendHeartbeatInterval	int	Master 与 Slave 心跳包发送间隔，默认为 5s
haHousekeepingInterval	int	Master 与 Slave 长连接空闲时间，超过该时间将关闭连接
haTransferBatchSize	int	一次 HA 主从同步传输的最大字节长度，默认为 32K
haMasterAddress	int	Master 服务器 IP 地址与端口号
haSlaveFallbehindMax	int	允许从服务器落户的最大偏移字节数，默认为 256M。超过该值则表示该 Slave 不可用
brokerRole	enum	Broker 角色，分为 ASYNC_MASTER,SYNC_MASTER, SLAVE，默认为异步 Master(ASYNC_MASTER)
flushDiskType	enum	刷盘方式，默认为 ASYNC_FLUSH(异步刷盘)，可选值：SYNC_FLUSH(同步刷盘)
syncFlushTimeout	int	同步刷盘超时时间

（续）

参数名	参数类型	描述
messageDelayLevel	String	延迟队列等级，默认"1s 5s 10s 30s 1m 2m 3m 4m 5m 6m 7m 8m 9m 10m 20m 30m 1h 2h"
flushDelayOffsetInterval	long	延迟队列拉取进度刷盘间隔。默认 10s
cleanFileForciblyEnable	boolean	是否支持强行删除过期文件。默认为 true
warmMapedFileEnable	boolean	是否温和地使用 MappedFile，默认为 false，如果为 true，将不强制将内存映射文件锁定在内存中
offsetCheckInSlave	boolean	从服务器是否坚持 offset 检测，默认 false
debugLockEnable	boolean	是否支持 PutMessage Lock 锁打印信息，默认为 false
duplicationEnable	boolean	是否允许重复复制，默认为 false
diskFallRecorded	boolean	是否统计磁盘的使用情况，默认为 true
osPageCacheBusyTimeOutMills	long	putMessage 锁占用超过该时间，表示 PageCache 忙，默认为 1s
defaultQueryMaxNum	int	查询消息默认返回条数，默认为 32
transientStorePoolEnable	boolean	Commitlog 是否开启 transientStorePool 机制，默认为 false
transientStorePoolSize	int	transientStorePool 中缓存 ByteBuffer 个数，默认 5 个
fastFailIfNoBufferInStorePool	boolean	从 transientStorePool 中获取 ByteBuffer 是否支持快速失败，默认为 false

表 A-5 FilterServer 配置属性

参数名	参数类型	描述
rocketmqHome	String	RocketMQ 主目录，默认为用户主目录
namesrvAddr	String	NameServer 地址
connectWhichBroker	String	FilterServer 连接的 Broker 地址
filterServerIP	String	FilterServer IP 地址，默认为本地服务器 IP
compressMsgBodyOverHowmuch	int	如果消息 Body 超过该值则启用
zipCompressLevel	int	Zip 压缩方式，默认为 5，详细定义请参考 java.util.Deflater 中的定义
clientUploadFilterClassEnable	boolean	是否支持客户端上传 FilterClass 代码。默认为 true
filterClassRepertoryUrl	String	filterClass 服务地址，如果 clientUploadFilterClassEnable 为 false，则需要提供一个地址从该服务器获取过滤类的代码
fsServerAsyncSemaphoreValue	int	FilterServer 异步请求并发度，默认为 2048
fsServerCallbackExecutorThreads	int	处理回调任务的线程池数量，默认为 64
fsServerWorkerThreads	int	远程服务调用线程池数量，默认为 64

推荐阅读

推荐阅读

推荐阅读

RocketMQ实战与原理解析
作者：杨开元

本书由云栖社区官方出品。

作者是阿里资深数据专家，对RocketMQ有深入的研究，并有大量的实践经验。在写这本书之前，作者不仅系统、深入地阅读了RocketMQ的源代码，而且还向RocketMQ的官方开发团队深入了解了它的诸多设计细节。作者结合自己多年使用RocketMQ的经验，从开发和运维两个维度，给出了大部分场景下的优秀实践，能帮助读者在学会使用和用好RocketMQ的同时，尽量少"踩坑"。同时，本书也结合源码分析了分布式消息队列的原理，使读者可以在复杂业务场景下定制有特殊功能的消息队列。

RocketMQ技术内幕
作者：丁威 周继锋

本书由RocketMQ社区早期的布道者和技术专家撰写，Apache RocketMQ创始人/Linux OpenMessaging创始人兼主席/Alibaba Messaging开源技术负责人冯嘉的高度评价并作序推荐。

源码角度，本书对RocketMQ的核心技术架构，以及消息发送、消息存储、消息消费、消息过滤、顺序消息、主从同步(HA)、事务消息等主要功能模块的实现原理进行了深入分析，同时展示了源码阅读的相关技巧；应用层面，本书总结了大量RocketMQ的使用技巧。通过本书，读者将深入理解消息中间件和底层网络通讯机制的核心知识点。